T0137040

Mining Over Air: Wireless Communication Networks Analytics

Ye Ouyang • Mantian Hu • Alexis Huet
Zhongyuan Li

Mining Over Air: Wireless Communication Networks Analytics

 Springer

Ye Ouyang
Verizon Wireless
Basking Ridge, NJ, USA

Mantian Hu
Chinese University of Hong Kong
Shatin, Hong Kong

Alexis Huet
University of Lyon
Lyon, France

Zhongyuan Li
Verizon Wireless
Basking Ridge, NJ, USA

ISBN 978-3-030-06403-7   ISBN 978-3-319-92312-3   (eBook)
https://doi.org/10.1007/978-3-319-92312-3

Printed on acid-free paper

This Springer imprint is published by the registered company Springer International Publishing AG part of
Springer Nature.
The registered company address is: Gewerbestrasse 11, 6330 Cham, Switzerland

With the warmest love to my daughter Evelyn Linlang Ouyang and my wife Rongrong Xu, for being always there with endless support.

Dr. Ye Ouyang

I want to thank my parents, Xing Huang and Xuefeng Hu, for always being there for me.

Dr. Mantian Hu

I would like to thank my wife, Minjoo, and my son, Daniel (Yul). You are the ones who keep me grounded and help me reach for the stars!

Zhongyuan Li

Contents

Chapter 1
Introduction

1.1 Big Data Analytics in Telecommunication Industry

The telecommunications ecosystem is a naturally big data warehouse, which contains a treasure trove of intelligence for those who know how to mine it. However, "big data" is not the answer to making sense of this data. Why? Because big data analytics for telecommunications is not a database problem; it is a problem of understanding the telecom data. Thanks to the evolution of networks and proliferation of smart phones, Communications Service Providers (CSPs) have access to huge amounts of subscriber, network, and application data, which are a tremendously valuable asset of information. Also, thanks to the power of big data analytics, the communications service providers (CSPs) can uncover important insights into network patterns and consumer behaviors.

Since birth, big data is made for the telecommunication industry. While talking about big data, the telecom industry has a unique advantage due to the absolute breadth and depth of data it collects in the course of normal business. Every day the telecommunication operators live in the world of big data. Big data has become a ubiquitous part of telecom industry because of the huge amount of data being generated every second through the connected world: when a subscriber makes a voice, video, or data call, sends a text message, surfs the Internet, etc.

The telecom world has seen exponential data growth in the last few years. The advent of smartphones, mobile broadband, internet of things, and 5G, etc. have all contributed to a huge volume of data. This has brought numerous and unpredictable changes to telecommunication network ecosystems, such as much heavier signaling traffic, concurrent connections of new applications, and changes in the data traffic consumed by every data application connection. The result of all those facts is a significant increase of data usage as well as explosive bandwidth consumption [1].

The growth rate of dense integrated circuit technology, first noticed by Intel's Gordon E. Moore and known commonly as Moore's law, appears to apply even more clearly in analysis of data usage users create and transmit in telecommunication

© Springer International Publishing AG, part of Springer Nature 2018
Y. Ouyang et al., *Mining Over Air: Wireless Communication Networks Analytics*,
https://doi.org/10.1007/978-3-319-92312-3_1

networks. In recent years there has been exponential growth in 4G users across the world. Global LTE subscriptions reached a total of 2.1 billion by Q1 2017 [2]. Such staggering growth also drives network traffic to rise quickly. Global mobile data traffic grew 63% in 2016. Global mobile data traffic reached 7.2 exabytes per month at the end of 2016, up from 4.4 exabytes per month at the end of 2015. Mobile data traffic globally has grown 18-fold over the past 5 years [3]. Machine-to-machine (M2M) communications is expected to overtake human generated data in 5G. Projections are of 32 billion devices generating 44 trillion GB of data by 2020 [4].

With the snowballing effect of the data growth generated from devices, networks, applications, and services, the analytics of telecommunications becomes crucial for the CSPs to truly understand networks, customers, business, and the industry itself. Most CSPs have been motivated to leverage data analytics for boosting the efficiency of their networks, segment customers, and drive profitability with some success. Over the years, CSPs actually have used a variety of techniques to work with these data including statistical analysis, data mining, knowledge management, and business intelligence. When captured wisely and analyzed professionally, the massive amount of data can reveal powerful insights to boost internal efficiency. Moving forward, big data analytics for the telecom industry poses a more aggressive challenge: how to gain deeper understanding, uncover insights, patterns and correlations, discover meaningful information from mounds of data, and finally take insightful actions to increase revenues and profits across the entire telecom value chain (from network operations to product development to marketing, sales, and customer service - and even to monetize the data itself) [5]. The CSPs can even leverage the insights to help other industries such as agriculture, power utilities, and health care, to name a few.

Communications service providers (CSPs) are at the heart of the telecom big data universe. They are sitting on a gold mine of digital data that enables them to understand their networks, services, and subscribers at an unparalleled level. Big data is much needed in a competitive landscape where over-the-top (OTT) players, such as Google and Facebook etc. are eating into their revenues. Buffeted by the growing thread from those non-traditional rivals, pressure to reduce costs, drifting customer loyalties, and a dynamic technological landscape, big data provides a unique opportunity for communications service providers (CSPs) to become more competitive and reverse recent declines in revenue and profit [6].

1.2   Driving Forces of Telecom Big Data Analytics

The entire telecommunication industry is living in an increasingly tough environment with fierce competition, network neutrality, direct threats from OTT players and technology transformation under the background of convergence of telecom and IT. CSPs need to venture out of comfort zones and avoid being outmaneuvered by non-traditional players. Accordingly, the driving force to fully leverage the power of big data in telecom industry is obvious.

The first driving force is that the telecommunication industry is in an increasingly tough environment with intense market competition resulting in reduced margins and declining Average Revenue per User (ARPU). Can a new business model be established through telecom data analytics to overturn the decline trend? Monetizing the telecom data through analytics can be an option.

Secondly, the OTT players such as Facebook, Google, Snapchat, Netflix, etc. pay nothing to use the operator's networks. On top of the free networks, OTT players provide voice, data, and content services to customers, due to which it directly impact telecom operator's revenue significantly. Network neutrality policy enables OTT companies to drive free on the highway paved by CSPs with huge CAPEX/OPEX, due to which APRU from traditional voice and data services are steadily declining.

Thirdly, the telecommunication companies need to keep pace with growing technological transformation in the convergence of IT and telecommunications (ICT). CSPs have a unique strength of owning the full stack data from physical layer to application layer compared to OTT players which owns only application layer data. With the right analytics solutions and products, CSPs will have more holistic view and insights on the entire ICT industry, which help CSPs overturns the competition against OTT players.

Fourth, following the last global economic decline, the telecommunication industry is not an exception at all. All the players in the telecom industry have been under immense pressure to improve operational efficiencies and reduce costs while maintaining quality of service at an optimal level. Telecommunication analytics also want to target two areas: improve internal business efficiency and data monetization with a new business model. Most CSPs have started to leverage analytics applied in their internal data to boost the operational efficiency of their networks, devices, services, applications, customers, and drive efficiency with some success. However, the potential of big data can be more interesting: expanding the analytics with much larger amounts of information for monetization across the entire telecommunication value chain, from network operations to new product development, marketing and sales, customer care, and even to monetizing the telecom data itself for other industries.

1.3 Benefits of Big Data Analytics for Value Chain of Telecom Industry

Big data analytics for CSPs acts as the vehicle toward insights. Data analytics in the telecom sector involves analytical methods, analytics tools, analytics technology, telecommunications domain knowledge, and subject matter experts (SMEs) in both data science and the telecommunication space. This is all required to make correlations between data points, identify trends, patterns and predict outcomes. It lays the foundation for the telecom businesses to start asking the right questions that lead to the holy grail of big data: insights.

For example, if a CSP wants to improve on its customer services at retail stores, it could collect data on customer sentiment through a host of contact points. Subsequent analysis of this data will reveal information about the CSP's customers, the services they prefer, frequency of services used, overall brand sentiment, and so on. While extremely valuable, these outputs are only the starting block for a deeper journey into how the CSP can optimize its services to cater to the needs of its legacy and emerging market. So, while analytics clearly facilitates a deeper understanding of customer demographics and sentiment, it will be up to the CSP to scrutinize their findings in a way that will lead to consequential insights.

How much can telecommunication industry benefit from big data? It is a critical question.

First, networks must be understood better. Network analytics enables the MNOs to take advantages of the network information within the networks for MNOs to operate the networks more reliable, robust, and scalable. Analytics benefits the MNOs throughout the entire life cycle of networks: network planning, network deployment, and network maintenance (optimization). In network planning phase, analytics first prepares the networks for future demands. Network planning analytics helps the MNOs understand the future demands of the network traffic in the network dimensioning. The CAPEX invested on the new network infrastructure or network expansion can be well planned in advance. In network deployment and optimization phase, MNOs can fully leverage the network analytics for optimizing their network performance and quality through analytical diagnostics methodologies.

Secondly, understand the customer better. Capabilities of big data makes it easier to understand the customer's profile, behaviors, and patterns from network, device, application, and social media data information, which further helps establish user driven performance indicators that will further enable the MNOs to understand user quality of experience.

Third, understand the application better. Various OTT applications running over the networks bring numerous and unpredictable changes to wireless networks, such as much heavier signaling traffic, concurrent connections of new applications, and changes in the data traffic consumed by every data application connection. MNOs can leverage the analytics for better understanding how the applications are impacting its own networks and services. MNOs accordingly can uncover important insights into application patterns and consumer behaviors.

1.4 Scope of Telecom Big Data Implementation

Compared with traditional data warehousing and database technologies in the telecommunication industry, big data can prepare networks for future demands and also understand the customers' quality of experience [7]. Specifically, big data helps businesses take advantage of the potential information and data within their networks in order to make them robust, optimized, and scalable. It can help optimize routing and quality of service by analyzing network traffic or patterns in real time.

Big data, with its capabilities, makes it easier to understand customers in detail right from network data and also social media information, which further helps establish customer-centric KPIs which enables to understand user experience. In this chapter, we will introduce the big data technology, use cases on top of it, the most state-of-the-art research and challenges by using big data technologies in network analytics, network customer & market analytics and business models.

1.4.1 Network Analytics

Mobile operators need to visualize their networks to understand how the network serves both for the internal management and external customers. Some failed data collection from the eNB can cause service degradation or outage. Replacement of equipment is usually more expensive than repair. An optimal schedule is needed for maintenance. Nowadays, telecommunication industry is migrating from traditional hardware and appliance-centric deployments to cloud-based deployments. In such cloud-based network deployment, the critical component of all network functionality – NFV (Network Function Visualization) or SDN (Software Defined Networking) [8, 9] are developed. Both of these are targeted to visualize network applications as well as the network connectivity. Big data analytics tools save unstructured, streaming and sensor data from networks. Among the current big data tools, Hadoop or Spark platform stores and processes unstructured, streaming, sensor data from the network. Mobile operators derive optimal maintenance schedules by comparing real-time information with historical data. By using MLlib or ML (which is a high level API) in spark or other machine learning libraries, algorithm can help mobile operators analyze their network, hence to reduce both maintenance costs and service disruptions by fixing equipment before it breaks.

1.4.1.1 Call Drop Analytics

Mobile operators expand broadband services and also focus on scaling up their network performance [10]. Disruption or outage in network can make a call drops and cause a poor voice service quality. Such event can harm the reputation of the telecom provider and can also increase the attrition among its customers. Hence, mobile operators should continuously monitor their networks for such disruptions and also resolve root causes at the very early stages. Customers may not always report call drops but would have a greater propensity to churn out in search of better services/coverage.

To address such issues, mobile operators can analyze CDR data generated by customers, and correlate with corresponding time interval network device logs, and then classify the reasons for call drops. In Hadoop big data platform, Flume it the tool to handle the data importing, and has the capability to ingest millions of call detail records into Hadoop. In a real time mechanism, Apache Storm uses this data to run pattern recognition algorithms which identify any troubling patterns.

1.4.1.2 Anomaly Detection

Anomalies in wireless networks are considered as unusual traffic patterns that deviate from the normal network behaviors [11]. Anomalies are referred as abnormalities, deviants, or outliers in data mining and machine learning. These anomalies in wireless network can be root caused by a variety of issues such as implementation of new features, network intrusions or disaster events. In many cases, intrusion for example, the outliers can only be discovered as a sequence of multiple data points, rather than as an individual data point. With more powerful capacity, network monitoring devices in recent years are able to collect data with very high sampling rate.

With big data platform, relevant information from a large amount of noisy data can be extracted via a designed effective anomaly detection system. In most applications, the collected data is generated by more than one process, namely co-occurrence data. Co-occurrence data are joint occurrences of pairs of elementary observations from two sets: traffic data (observations-W) in one set are associated with the generating entities (Time stamp or node ID-D) in the other set. Modeling co-occurrence data (traffic data with generating entity) is a fundamental problem in anomaly detection. It poses a challenge for effective anomaly detection when the usual distribution varies with generating entities (time slot or node ID).

With the help of machine learning libraries like MLlib in Spark, such patterns or outliers of the network behaviors could be detected and identified in an efficient method.

1.4.1.3 Network Performance Healthiness

Workflow of traditional network optimization follows a few routine steps. A network system performance engineer normally first pulls the KPI stats from Operation Support System (OSS) tool, eyeballs the raw data, and visualizes the KPI trend. The engineer leverages the domain knowledge or some handcrafted rules such as KPI threshold to look for unusual patterns, anomalies, and/or killer KPIs. After locking down the issue, the engineer needs to identify the root causes of the symptoms in terms of service degradation, coverage/capacity black holes, heavy hitters/users, capacity bottlenecks etc. Meanwhile the engineer also looks into the network remedy tickets if any to validate the performance issues. After walking through all the steps above, the engineer finally concludes a solution based upon all the information he/she collected and analyzed. The solution is made combing his/her own domain knowledge, experience, along with some half-automated solutions through some network optimization tools he/she uses.

Obviously the engineers get lost in eyeballing thousands of KPIs in evaluating network performance every day. There should be an AI like scientific method, such as tree or neuron alike models, from top of down in a divide and conquer manner to filter out noisy data and less important information and navigate the engineers to focus on the critical KPIs. With this way the engineers will be free from the tedious eyeballing task and more concentrate on performance diagnostics and optimization, which is the most valuable step in network optimization. The NPH (Network Performance Healthiness) can be defined and calculated to evaluate and visualize the performance of network at different level, like cell, eNodeB, or a pre-defined Geo-bin on top of a big data platform.

1.4.1.4   Intelligent Network Planning

Mobile operators need network planning solutions with advanced analytics to federate and correlate information from multiple network data repositories. It provides operators with abilities to make plan, predict, and optimize their investment in network, and also provides the prioritized and optimal network investment plan based on service forecast demands. Network planning systems must be advanced analytics-driven and work closely with their OSS systems. Such systems can drive capacity optimization and provide network planners with ability to create "what if" scenarios.

1.4.1.5   Cell-Site Optimization

4G and future 5G networks are intended to implement SON (Self-Organizing Network) functionalities. One of the most important functions in SON is self-optimization where cells automatically managing how they interact with one another, managing their power consumption and how they load balance traffic and handover traffic between cells. Such functions depends on if mobile operators can augment the network performance with contextual information. This function includes subscriber information such as user experience in specific areas, how that user experience varies according to the different types.

1.4.1.6   Subscriber-Centric Wireless Offload

Applications can collect data from remote cell site monitoring solutions, DPI systems, customer usage systems, backhaul network management systems. Big data and machine learning technologies can be utilized with such application to process and analyze the data with big volume. Subscriber data repositories can be used in real time to push different types of traffic belonging to different types of customers to different cells. Since WiFi is implemented to offload traffic with 4G network, contextual intelligence tools can correlate customer information with their lifetime value and intelligently decide which subscribers should be offloaded on WiFi.

1.4.1.7 Congestion Control

RAN (Radio Access Network) congestion is one of the main problems for mobile operators. With subscriber information, services and location information, we can provide visibility at cell level and provide priority to subscribers. Because congestion events are fleeting, it is a key area where mobile operators are planning to leverage big data to identify the problems and deploy RAN congestions.

1.4.2 Customer & Market Analytics

1.4.2.1 Churn Prediction

Retaining customers is one of the most important tasks or challenges for any mobile operators. Churn is defined as the process of prediction of customers who are at risk to leave a mobile operator. Acquiring new customers is more expensive than retaining old users.

   With the help of predictive models and machine learning algorithms, we are able to accurately identify mobile customers who are likely to stop the service. With the data of customer usage, complaints, transactions, social media, algorithms can create factors to identify customers moving out.

1.4.2.2 Customer Profiling

Customer profiling is the process of splitting market or customers into different groups according their behave similarity. Such analytics is very important for mobile operators. For example, mobile operators can create tailored products for customers, identify high-value and long-term customers, and identify potential customers by analyzing customer value segmentation.

   With such functionalities, operators could identify loyal customers who have a high potential lifetime value, or enable targeted marketing and retention activities to reduce the churn rates. Using big data technologies, operator can utilizes predictive models to helps them to determine customers that are more likely to repeat purchases/buying patterns.

1.4.2.3 Predictive Campaign & Preemptive Customer Care

Nowadays, mobile operators are facing challenges to retain their mobile customers. Real-time analytics technologies could help operators to analyze visualize and predict the customer churn and revenue losses in a network. Data analyzed from the real-time data could uncover purchasing patterns. Such patterns are highly-personal, and they drive responses that are ultra-timely. Mobile customers are able to get what they want, when they want it about the customers. At the same time, businesses could get real-time data and analyze them to learn which makes future offerings more targeted.

With big data technologies, mobile operators can get high volume of campaigns tools. Such tools or methods encompass data management, campaign management and performance monitoring to handle volume of data which needs to be sifted through and to configure business.

1.4.2.4 Location-Based Services

With the geographic information, mobile operators can provide great insights of information to users. Such map based visualization can be used for many analytics services. Besides the location based service, the positioning technology using location technology can support the positioning service to user instead of WiFi-based positioning service.

1.4.3 Innovative Business Models

1.4.3.1 Data Exposure & API Enablement

For most of the application nowadays, application programming interface (API) provides the access method for operators to exchange internal and external data. With a well-defined API, application developers can connect the customer with a new application.

1.4.3.2 Using Payment Data for Increasing Sales

Filed communication payments data can be collected by the mobile operator and provided to the business customers. Such function can collect and analyze the customer payment and transaction data and send the sales plan or coupon to the users.

1.4.3.3 Math Demand & Offering Nearby

Mobile operator can meet requirements and demands from users in nearby areas, for example, shopping malls, supermarket, or restaurant, to provide customized service. Mobile operators could also share information with other business owners to match a list of subscribers. Therefore, customized information can be pushed to a specific group of users based on the subscription from business owner.

1.5   What This Book Covers

Chapter 1, *Introduction*, will give a complete overview of big data analytics in telecommunication industry.

Chapter 2, *Analytics for Telecom*, covers the machine learning algorithms which can be utilized in the analytics for telecommunication industry. This describes the algorithms of regression, classification, clustering, and forecasting methods. It also describes the ARIMA models and reinforcement learning.

Chapter 3, *Trending LTE Network Performance*, introduces the procedure of network performance analytics. For instances, Network Performance Forecasting Strategy, Relationship between network resources and performance indicators, forecasting network resources, and application for RRC connection setups evaluation.

Chapter 4, *Trending Device Readiness and Device Return Rate*, explains device return rate and readiness forecasting strategies, models and results of implementation.

Chapter 5, *Evaluating VoLTE Voice Quality in LTE network*, introduces definition, methodology and trail results of VoLTE voice quality in telecommunication network.

Chapter 6, *Profiling Wireless Resource Usage for Mobile Apps*, shows the tools, algorithms and experimental results of mobile resource management and usage.

Chapter 7, *Outlier Identification in Telecom Data*, introduces the outlier identification models and their comparison in telecommunication industry.

Chapter 8, *Self-Optimization in LTE Network*, focuses on the introduction of SON (Self-Organizing Network), APP-SON which is a self-optimization solution for 4G and future 5G network on top of big data platform, and its implementation.

Chapter 9, *Telecommunication Data and Marketing*, describes the telecom marketing, social network, and network measurement.

Chapter 10, *Contagious Churn*, concentrates on the churn problem in telecom industry and dynamic models for social learning and network effects.

Chapter 11, *Network Based Targeting*, introduces the channels of network effects, modeling strategies issues, and their findings and applications.

Chapter 12, *Social Influence and Dynamic Network Structure*, covers the models of network structure effects analytics on social influence.

References

1. Big Data & Advanced Analytics in Telecom: A Multi-Billion-Dollar Revenue Opportunity. https://www.huawei.com/ilink/en/download/HW_323807
2. Ericsson Mobility Report June 2017
3. Cisco Visual Networking Index: Global Mobile Data Traffic Forecast Update, 2016–2021 White Paper
4. Digital Cosmos to Include 32 Billion Devices Generating 44 Trillion GB of Data by 2020. http://www.industrytap.com/digital-cosmos-include-32-billion-devices-generating-44-trillion-gb-data-2020/28791

5. Benefiting from big data: A new approach for the telecom industry. https://www.strategyand. pwc.com/reports/benefiting-big-data

6. Analytics: Real-world use of big data in telecommunications. https://www-935.ibm.com/ser vices/us/gbs/thoughtleadership/big-data-telecom/

7. Hilbert, Martin. "Big data for development: A review of promises and challenges." Development Policy Review 34.1 (2016): 135–174.

8. Virtualization, Network Functions. "NETWORk FUNCTION VIRTUALIzATION."

9. Han, Bo, et al. "Network function virtualization: Challenges and opportunities for innovations." IEEE Communications Magazine 53.2 (2015): 90–97.

10. Lee, Jonghun. "Method and system for preventing call drop by restricting overhead message updated in 1X system during 1xEV-DO traffic state." U.S. Patent No. 7,394,787. 1 Jul. 2008.

11. Garcia-Teodoro, Pedro, et al. "Anomaly-based network intrusion detection: Techniques, systems and challenges." computers & security 28.1–2 (2009): 18–28.

Chapter 2
Methodologies of Analytics for Telecom

The last few years have witnessed a huge increase in the network flow of data, accompanied by a growing amount of data collected by mobile operators. Beyond storage and management of indicators about this flow, a main challenge has been to select and use this mass of material to get a better knowledge of the network. Indeed, since the resulting flow has become impossible to manually process and analyze, simple summarizing statistics of the network traffic have been insufficient to represent the full information contained in those data. This issue has led to the need of new strategies to manage and understand data.

A primordial set of tools built in the last decades is machine learning algorithms. They have been developed to provide advanced analytics that is to discover and interpret complex patterns in data. Those algorithms are normally categorized into two types: Supervised learning, which refers to techniques predicting outputs or classifying objects from a labeled training set; and unsupervised learning, which refers to techniques describing or segmenting objects to infer hidden structures of the data. In this chapter, a set of the most common machine learning algorithms are presented. Those algorithms have been selected because of their importance in the telecom field. Consequently, they will appear as a part of the methodology of the next chapters, and specifically applied for network analysis, evaluation or detection purposes.

This chapter comprises four main components: Regression, classification, clustering and forecasting methods. For each part, analytics algorithms are introduced, and focus is put on comprehension of the guiding concept. Detailed developments can be found in [1] (for regression, classification, clustering) and [2] (for forecasting).

© Springer International Publishing AG, part of Springer Nature 2018   13
Y. Ouyang et al., *Mining Over Air: Wireless Communication Networks Analytics*,
https://doi.org/10.1007/978-3-319-92312-3_2

2.1   Regression Methods

Regression methods are supervised methods for estimating relationships between predictors and a continuous target variable. Most commonly, regression analysis estimates the conditional expectation of the dependent variable given the independent variables. Less commonly, the focus is on a quantile, or other location parameter of the conditional distribution of the dependent variable given the independent variables. In all cases, a function of the independent variables called the regression function is to be estimated. In regression analysis, it is also of interest to characterize the variation of the dependent variable around the prediction of the regression function using a probability distribution. Regression analysis is widely used for prediction and forecasting, where its use has substantial overlap with the field of machine learning. Regression analysis is also used to understand which among the independent variables are related to the dependent variable, and to explore the forms of these relationships. In restricted circumstances, regression analysis can be used to infer causal relationships between the independent and dependent variables.

In the context of telecom, some variables may be easier to collect than others and the relationship between collected variables and Key Performance Indicators (KPI) may remain unclear due to unknown or uncontrolled internal or external factors. In this way, the relationship between predictors and the target variable can be shaped through probabilistic models to understand and control this uncertainty. Those methods help to predict value of the target variable for a new set of predictors and to build confidence of interval around this prediction. In this section, a brief presentation of common regression methods is done, in order to give key elements for understanding the following chapters. The most simple regression method is the linear regression and is defined in Sect. 2.1.1 (see also [3] for a brief history of the linear regression). LOESS [4] (stands for *local regression*) and generalized additive model [5] (GAM) methods are then defined in Sect. 2.1.2 and give a flexible way to deduce non-linear relationships between predictors and the target variable. The concept of regularization by automatically selecting features is finally introduced in Sect. 2.1.3 through the Lasso regression [6].

When the target variable is categorical, we instead use supervised classification methods (as shown in Sect. 2.2) to study relationships between predictors and this target variable.

2.1.1   Linear Regression

Linear regression is a linear approach for modeling the relationship between a scalar dependent variable y and one or more explanatory variables (or independent variables) denoted X. The case of one explanatory variable is called simple linear regression. For more than one explanatory variable, the process is called multiple linear regression. Linear regression was the first type of regression analysis to be studied rigorously, and to be used extensively in practical applications. This is because models which depend linearly on their unknown parameters are easier to fit than models which are non-linearly related to their parameters and because the statistical properties of the resulting estimators are easier to determine.

In linear regression, the relationships are modeled using linear predictor functions whose unknown model parameters are estimated from the data. Such models are called linear models [3]. Most commonly, the conditional mean of y given the value of X is assumed to be an affine function of X; less commonly, the median or some other quantile of the conditional distribution of y given X is expressed as a linear function of X. Like all forms of regression analysis, linear regression focuses on the conditional probability distribution of y given X, rather than on the joint probability distribution of y and X, which is the domain of multivariate analysis. In the classic linear regression, relationship between the target variable y and the predictors $x_1, \ldots,$ x_l is assumed to be linear. In the most classic form, it is then linked with the following equation:

$$y = \beta_0 + \beta_1 x_1 + \ldots + \beta_l x_l + \varepsilon,$$

where predictors are fixed known values, $\beta_0, \ldots, \beta_l$ are fixed unknown parameters we wish to fit, and ε is a random variable following a Gaussian distribution with mean 0 and variance σ^2. In addition, we assume that for all pairs of data instances i and j, the related random variables ε_i and ε_j are independent. The response variable y therefore follows a Gaussian distribution.

The usual algorithm computed to fit parameters is the least squares method, and a direct matrix calculation gives those coefficients (details can be found in [7]). In this regression method, all predictors contribute to the final prediction, positively or negatively. An example of linear fit is performed in Fig. 2.1 (a) with one predictor.

As shown in Fig. 2.1 (a), the direct linear regression is limited to accurately fit complex relationships between predictors and the target variable. The polynomial regression is a direct extension of the linear regression which is able to fit polynomials under a fixed degree D. Instead of taking $x_1, \ldots, x_l$ directly, we first compute power values $x_1^d, \ldots, x_l^d$ for d from 1 to D, leading to Dl features. The model has then $Dl + 1$ coefficients to fit (related to the Dl features plus the constant coefficient), instead of $l + 1$ in the classic linear model. The fit is performed as before with the least squares method. An example of polynomial fitting with degree up to 3 is shown in Fig. 2.1 (b).

2.1.2 Non-linear Regressions

Nonlinear regression is a form of regression analysis in which observational data are modeled by a function which is a nonlinear combination of the model parameters and depends on one or more independent variables. The data are fitted by a method

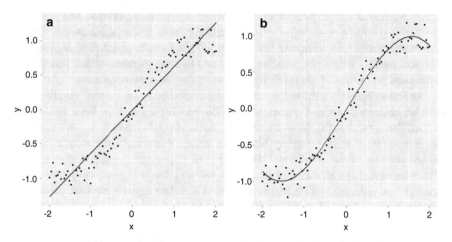

Fig. 2.1 Two regression algorithms applied to deduce a relationship between x and y. 100 points x are sampled on the interval $[-2, 2]$. The relationship between x and y is defined by $y = \sin(x) + \varepsilon$, where each ε follows a centered Gaussian distribution with variance $\sigma^2 = 1/100$. (**a**) Regression with linear regression. We observe that the fit can capture the main trend of the relationship between x and y. However, the capacity of this algorithm is too low to capture non-linear changes. (**b**) Regression with a polynomial regression of degree 3. For this fit, the whole trend is captured although we fit a sinusoidal function with a polynom. One may note that extrapolation of this fit will eventually lead to very poor results

of successive approximations. Non-linear regression methods are useful to provide more flexibility in the fitting, at the price of lack of understanding of the underlying model. The capacity of those methods can be very high and can lead to overfitting if the hyperparameters are not correctly selected (see [8] chapter 5 for a detailed discussion of capacity and overfitting).

The most common non-linear regression is LOESS [4], which is also a non-parametric regression. The idea of the method is to perform local regression around each point of interest and can be described as follows: To fit a new point given the predictors, a subset of the data around this point is selected. Then, a low-degree polynomial fit of this subset is performed (by giving more weight to observed values close to the point of interest). The fit for the point of interest is then defined as the LOESS fit for this point. As we can show, two hyperparameters have to be selected: The degree of the polynomial fit, and the percentage of the total set selected. In general, the degree if set to 1 or 2, an higher number causing overfitting and unstable behavior issues. The percentage of the total set selected can be changed to allow the fit to be more or less smooth (the higher the smoother).

An example of the interest of LOESS compared to fixed parametrized models is illustrated in Fig. 2.2. In (a), a regression with a polynomial regression of degree 3 is performed, but the regression has not the flexibility to retrieve the whole behavior of the curve (this can be shown around $x = 1.5$, where the fitted curve stands below the scatterplot). In (b), the regression using LOESS can correctly fit the scatterplot and can be used to successively predict new points located in the interval $[-2, 4]$.

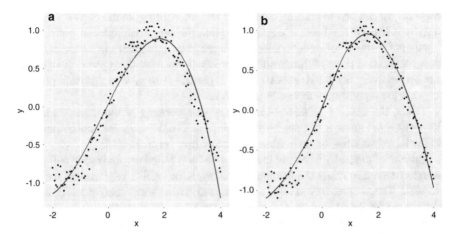

Fig. 2.2 Two regression algorithms applied to deduce a relationship between x and y. 150 points x are sampled on the interval $[-2, 4]$. The relationship between x and y is similar to the one described in Fig. 2.1. (**a**) Regression with a polynomial regression of degree 3. Around $x = 1.5$, the fitted curve stands below the scatterplot. (**b**) Regression with LOESS (automatically fitted with *ggplot2* package in R). We observe the fitted line correctly stays at the middle of the scatterplot and can be used to successively predict new points located in $[-2,4]$

Another more complex method to perform flexible non-linear regression is GAM [5, 9]. Compared to LOESS, it performs better as soon as there are a sufficient amount of observations in the data set (for example from 1000 elements). In GAM, relationships between target variable y and predictors $x_1, \ldots, x_l$ are linked with the following equation:

$$y = \beta_0 + f_1(x_1) + \ldots + f_l(x_l) + \varepsilon,$$

where $f_1, \ldots, f_l$ stand for non-linear links between input variables, β_0 is the constant term, and ε follows a Gaussian distribution. Functions f_j can be estimated with a non-parametric backfitting algorithm. This algorithm is iterative and at each step, functions f_j are approximated using cubic third splines.

Other examples of nonlinear functions include exponential functions, logarithmic functions, trigonometric functions, power functions, Gaussian function, and Lorenz curves.

2.1.3 Feature Selection

Feature selection is also known as feature engineering. Feature engineering is the process of using domain knowledge of the data to create features that make machine learning algorithms work. Feature engineering is fundamental to the application of machine learning, and is both difficult and expensive. The need for manual feature engineering can be obviated by automated feature learning. In the Figs. 2.1 and 2.2, only one feature has been selected to explain the target variable. In practice, the number of features collected in telecom is large (tens or hundreds of variables), and a new issue called curse of dimensionality arises in this case: the regression algorithm can successively fit a fixed set but will fail to generalize to new unseen examples. We say that the algorithm with overfit in the case.

Often, the number of features necessary to accurately fit the target variable is lower than the number of features available, since some variables may be unrelated to the target variable, or since some predictors may be correlated.

Instead of manually selecting features, a method called regularization can help to weight or even to select the features of interest which effectively explain the target variable. The simpler way of regularization is to restrict some coefficients (known as shrinkage), by penalizing them if there are too large.

In the case of the linear regression, the constraint concerns the coefficients $\beta_0, \ldots, \beta_l$. The Least Absolute Shrinkage and Selection Operator [6] (LASSO) regression imposes a condition on the absolute sum of coefficients (excepted the constant coefficient), as written in the following equation:

$$\sum_{k=1}^{l} |\beta_k| \leq t,$$

where t is a parameter to select. Theoretical considerations show that this is an hard constraint on coefficient, which means that some of them can be exactly equal to zero. This method is therefore able to do feature selection of predictors. Ridge regression [10] has some similarities with LASSO regression but with a regular constraint function (by constraining the sum of squares instead of absolute values):

$$\sum_{k=1}^{l} \beta_k^2 \leq t.$$

Ridge regression is able to shrink coefficients and is also widely used to regularize the fit. Finally, the elastic net regularization regression [11] gives another way to shrink elements, by making a compromise between the ridge and the Lasso regression. The constraint function is here a linear combination between two previous regularizations:

$$\alpha \sum_{k=1}^{l} \beta_k^2 + (1 - \alpha) \sum_{k=1}^{l} |\beta_k| \leq t.$$

Other methods to select features also exist, for example by selecting iteratively the features which are able to explain the target variable (subset selection in [12], univariate filters method in [13]).

Besides the above feature engineering methods, another efficient method is principal component analysis (PCA). PCA is designed to reduce the dimensionality of features which are correlated with each other in high-dimensional dataset. PCA maps data from high dimensional feature space into low dimensional representation by calculating the correlation among features. After mapping, all of the data points can be represented by the principle orthogonal components in a low dimensional feature space. Among all ordered components, the first component is considered as the one retaining maximum information from the original features. In the low dimensional feature space, the first component lies on the first coordinate, the second one the second, and so on [14]. For example, each row in a matrix X represents a data point, and each column in the matrix represents different features. The PCA transformation is defined by a set of p-dimensional vectors of weights:

$$\mathbf{w}_{(k)} = \left(\varpi_{(1)}, \varpi_{(2)}, \ldots, \varpi_{(p)}\right)_{(k)}$$

The p-dimensional vectors in X can be mapped to new vectors with principle component scores:

$$\mathbf{t}_{(i)} = \left(\mathbf{t}_{(1)}, \mathbf{t}_{(2)}, \ldots, \mathbf{t}_{(m)}\right)_{(i)}$$

Where $t_{(k,i)} = x_{(i)} \cdot w_{(k)}$, for $i=1,\ldots,n$ and $k=1,\ldots m$. t is defined as the maximum possible variance from x.

With PCA, the extracted components can help to improve the efficiency in the training process and also reduce the calculation complexity. In this algorithm, the process of feature transformation is not simply dropping features. Each principle components is a combinational result calculated from the original features.

2.2 Classification Methods

In some applications and contrary to Sect. 2.1, we may want to study between predictors $x_1, \ldots, x_l$ and a categorical target variable y. Because the target variable is categorical, regression methods are not able to correctly fit and predict this variable. New tools have been developed for this type of data and known as classification methods. In this introduction of classification methods, we will focus on the $0 - 1$ classification: we assume that the target variable can only take value 1 (for example corresponding to a success) or 0 (for example corresponding to a failure). The classification task must then understand how the predictors are related with the target variable, and usually output a probability of success of the output.

The main method described in this section is the logistic regression [15, 16] (Sect. 2.2.1). This basic method, which can be seen as an extension of the linear regression model, is still often used despite its simplicity. Many other classification algorithms exist and are able to manage nonlinear patterns. A short description of the main algorithms is given in Sect. 2.2.2.

2.2.1 Logistic Regression

The name *logistic regression* is somewhat poorly-named since it is a classification
method instead of a regression method. This method is usually dedicated to perform
binary classification. As a linear classifier, it cannot catch complex non-linear
patterns. Also, it is sensitive to correlation within predictive variables. Thus, correlations have to be checked to avoid over-fitting and over-confidence in certain
variables. Its main advantage is to be quick to run and reliable for linear patterns
identification. Furthermore, as white box model, impact of each feature over the
target variable is easy to understand.

Contrary to the linear regression where the output can take value on the real line,
the logistic regression constraint output values to be between 0 and 1. By doing this,
the fitted results can be seen as the probability for the target to effectively be equal to
1. The way to constraint values from the real line to $(0, 1)$ is mapped with the sigmoid
function, defined as follows (and plotted in Fig. 2.3):

$$\sigma(x) = \frac{1}{1 + \exp(-x)}.$$

Therefore, the logistic model can be written as:

$$p(y = 1 | x, \beta_0, \ldots, \beta_d) = \sigma(\beta_0 + \beta_1 x_1 + \ldots + \beta_l x_l).$$

In this model, $\beta_0, \ldots, \beta_l$ are real parameters to optimize: given a training data set
where the target variable is available, we want to have $p(y = 1 | x, \beta_0, \ldots, \beta_d)$ close

Fig. 2.3 The logistic
function $x \mapsto \frac{1}{1+\exp(-x)}$

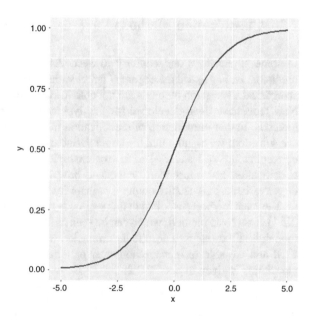

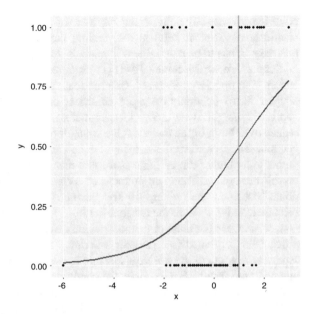

Fig. 2.4 Logistic regression applied on a sample data. The sample data is shown in black, the fitted line in blue and the separation between predicted $p(y|x, \beta) > 1/2$ and $p(y|x, \beta) < 1/2$ in orange

to 1 when the target variable is 1 and close to 0 when the target variable is 0. The most effective way to solve this optimization problem is by finding parameters maximizing the training data set likelihood. Since there is no closed-form solution of this problem (contrary to the linear regression), we use an iterative algorithm to approximate parameters, usually with a gradient-descent algorithm (details can be found in [17]).

In Fig. 2.4, we observe an example of logistic regression on sample data, with one predictor. The sample data (in black in the figure) show that low value (respectively high value) of the predictor seems related to a low probability (respectively high probability) to have $y = 1$. In this case, logistic regression can be applied and will linearly separate the predictor spate space. On the fitted line (in blue in the figure), we observe that the probability to have $y = 1$ increases with the predictor as expected. The separation (shown in orange in the figure) can successively classify the sample set with a low error rate.

2.2.2 Other Classification Methods

In this section, we briefly introduce other classification methods which are in use in the telecom field: k-nearest neighbors, Support Vector Machine (SVM) and decision trees.

The k-nearest neighbor's algorithm [18] is a non-parametric classification technique. To determine which class a new point belongs to, the k closest data points in the training data set are selected (via Euclidian distance). The prediction of the new point is performed by doing a majority vote of the k labeled candidates, since each candidate has a class label in the training dataset. In KNN algorithm, only the parameter k has to be determined. It controls the capacity of the learning method.

Support Vector Machine (SVM) [19] is a geometric method to perform classification. The main idea is to separate the feature space into two half-spaces (in its simpler form) by minimizing the empirical classification error and maximizing geometric margin. A regularization parameter is common to allow soft margin and reduce overfitting of the method. One strength of SVM is that it can be extended to get non-linear classifications, using the so-called *kernel trick*, by mapping data points to an higher dimensional space and then by performing linear classification on this new space. The main problem of SVM is it suffer from an large computational cost when the training set size becomes large, and is therefore not always suitable to solve tasks in the telecom field.

Decision tree is a class of methods which can perform classification according to a tree initially trained (see [20] for a review). Each node of the tree successively separates a region of the feature space into two sub-regions according to a condition on one of the feature. For example, the condition at the root node can be $x_1 \leq 2$, and in this case the sub-tree on the left is related to the half-space $x_1 \leq 2$ and the sub-tree on the right is related to $x_1 > 2$. Each node is then related to a region of the input space, this region being labeled by 0 or 1 according to the target variable. When a new point has to be fitted, we show in which region this point belongs to predict the target variable. Random Forest [21] is an extension based on computing a large quantity of decision trees, each tree being computed without considering earlier trees and which are then combined to deduce a single prediction. This approach has many advantage: It is less subject to overfitting compared to decision trees, the importance of each feature can be monitored and trees are relatively fast to train. For all these reasons, random forest has been a popular approach for classification from the 2000s'.

Random forests and gradient boosting trees [22] are two tree-based ensemble learning algorithms for classification. Both of them operate by constructing a multitude of decision trees at training time and leverage the power of multiple weaker classifiers. The difference between random forest and gradient boosting tree is that random forests is based on bagging and gradient boosting tree is build based on boosting.

2.3   Clustering Methods

An important class of machine learning algorithms applied in the telecom field is unsupervised learning [23, 25]. In this type of learning, the data are not labeled, and algorithms intend to learn the structure and to make sense of the data. The most common form of unsupervised algorithms is clustering algorithms, which divide an heterogeneous set of data into groups, where each group shares some similarities. One interest of such algorithms is to make sense of data to therefore detect unusual elements or group of elements outside a defined set.

For all the following methods, a set of data is available and we try to separate it into K subsets, where data instances contained into each subset share similarities (for example, close each other and far from the data instances in the other sets in the sense of the Euclidian distance). In this section, and commonly with clustering methods, the type of data is assumed to be numeric.

In the three first sections, we describe common clustering algorithms which are useful in the telecom field. We begin with the simple K-means algorithm [21] in Sect. 2.3.1, which is still popular today. Some drawbacks of K-means can be corrected with more advanced detection such as Gaussian Mixture Model [24] (Sect. 2.3.2). A large number of clustering algorithms exists for different purposes, and some of them are defined in Sect. 2.3.3. Finally, in Sect. 2.3.4, we remind some important requirements and experiments to use clustering methods successively.

2.3.1 K-Means Clustering

The basic example of clustering method is the K-means algorithm [21]. This algorithm separates the data set into a partition of K subsets, by trying to minimize the within-cluster sum of squares of each subset (which is a sum involving distance between each element of the subset and the mean element of this subset).

Explicitly, with $x_1, \ldots x_n$ points of $\mathbb{R}^d$, we try to find the partition $S_1, \ldots S_K$ minimizing

$$\sum_{i=1}^{K} \sum_{x \in S_i} \|x - \mu_i\|^2,$$

where for all i, μ_i is the mean of all elements of S_i.

This minimization problem cannot be solved exactly when the number of points contained in the data increase (for example, with 100 points and 2 clusters, the number of partitions to compare is 2^{100}). The Lloyd's algorithm allows to get an approximation of the exact solution. Usually, the solution founds with this algorithm is different from the exact solution, but is quite reliable for most purposes.

The algorithm is described as follows:

- Initialize the cluster centers by taking randomly K different points of the data set,
- Until convergence, follow those two steps:

  Assignment step: For each point, compute the distance between it and the cluster centers. Each point is then assigned the closest center. We obtain a partition of the set into K subsets.

  Updating step: Each cluster center is updated by taking the mean over all elements related with this center.

An example of this algorithm on sample data is given in Fig. 2.5.

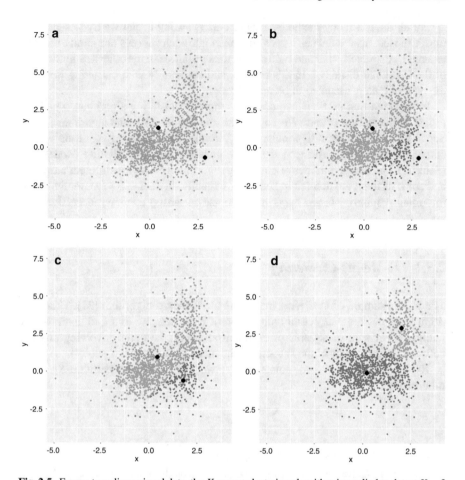

Fig. 2.5 From a two-dimensional data, the K-means clustering algorithm is applied to detect $K = 2$ clusters. The data are sampled from two Normal distributions: 1000 elements are sampled according to a distribution with mean $(0,0)$ and variance $\begin{pmatrix} 1 & 0 \\ 0 & 1 \end{pmatrix}$; 500 elements are sample according to a distribution with mean $(2,2)$ and variance $\begin{pmatrix} 1/4 & 0 \\ 0 & 4 \end{pmatrix}$. (a) Initialization: Two centers points (in black) are selected randomly among the available points (in gray). (b) Step 1 – Assignment: For each point, we compute the Euclidian distance between it and the center points. Each point is then assigned the closest center. We obtain a partition of the set into two subsets (in blue and red). (c) Step 1 – Updating: Each center is updated according to the elements of the data set related with this center. Explicitly, the new center is the mean over all elements related with this center. (d) Final step: After 15 iterations, the algorithm converges to a final partition of the data set. Compared with the parameters used for sampling, the resulting partition is fair and is able to help us representing structure of data

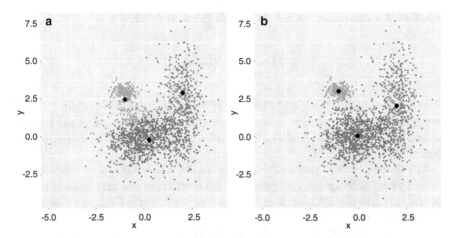

Fig. 2.6 Two clustering algorithm applied to detect $K = 3$ clusters. The data are sampled from three Normal distributions: The two firsts as done in Fig. 2.5, with in addition 200 elements sampled according to a distribution with mean $(-1,3)$ and variance $\begin{pmatrix} 9/100 & 0 \\ 0 & 9/100 \end{pmatrix}$. (**a**) Clustering with K-means: the model favorises even sized clusters, which tends to increase the size of the cluster shown in green. (**b**) Clustering with GMM (with *mclust* package in R): the cluster shown in green is correctly identified. In addition, variance is increased in the y direction for the cluster shown in blue

The K-means clustering algorithm only cares about minimizing the within-cluster sum of squares of each cluster. When the underlying variances of the clusters are different, or in the case of uneven sized clusters, the algorithm will fail to successively partition the data. An example of this behavior is shown in Fig. 2.5 (a). In this case, more advanced methods such as the Gaussian Mixture Model algorithm (described in the next section) can be used to perform clustering (Fig. 2.6).

2.3.2   Gaussian Mixture Model

The Gaussian Mixture Model [24] (GMM) assumes that all data points are generated from a mixture of K Gaussian distributions, each distribution parametrized by its mean and its matrix of variance-covariance. In addition, each cluster is linked to a fixed (initially unknown) probability to occur, so this method can manage uneven sized clusters.

We can formally write the probability of a point x of the set by decomposing it on all K possible clusters:

$$p(x) = \sum_{k=1}^{K} p(k)p(x|k).$$

The GMM assumes that we can rewrite this previous formula as:

$$p(x) = \sum_{k=1}^{K} \pi_k N(x|\mu_k, \Sigma_k),$$

where $(\pi_1, \ldots, \pi_K)$ are positive number such that $\sum_{k=1}^{K} \pi_k = 1$, where μ_k is the mean value and Σ_k the matrix of variance-covariance of the related cluster for each k, and where $N(x|\mu, \Sigma)$ stands for the density in x of a Normal distribution with parameters μ, Σ.

Estimation of the parameters of the model is usually done through an Expectation-Maximization [24] (EM) algorithm, which can be seen as an extension of the previous Lloyd's algorithm. The aim of the EM algorithm is to iteratively increase likelihood of the set by updating the parameters $(\pi_k, \mu_k, \Sigma_k)_k$ of the model.

Contrary to the K-means algorithm, GMM is able to cluster correctly more complex patterns, in the case where the Gaussian underlying assumption is reasonable. In Fig. 2.5, the GMM algorithm is compared to the K-means algorithm on a sample set. We observe that the dense set (around points shown in green) is correctly detected with GMM, which is not the case with K-means.

2.3.3 Other Clustering Algorithms

Many other clustering algorithms have been developed to fulfill some specific tasks. Some of them can be seen as extensions of the K-means algorithm and are not detailed here: The K-medoids [26] algorithm replaces means by so-called medoids during the updating step, which corresponds to take a point of the data set instead of simply the mean over the state space, and can be useful when strong anomalies are present in the data; The fuzzy C-means [27] algorithm assumes that data elements can belong to more than one cluster, which means that each element is related to a vector of size C representing the weight to be in each cluster; The CLARANS [28] (stands for Clustering Algorithm based on Randomized Search) algorithm only performs clustering on a subset of the whole data set and then assigns each element of the whole set to the closest labeled element. The CLARANS extension of K-means is useful to perform fast clustering computations for very large data sets.

Hierarchical clustering [28] is another class of clustering methods. In practice, this class is not usually used in the telecom field since it requires the computation of the distance matrix between each pair of element, which is computationally costly when the number of elements of the set is large. The idea of hierarchical clustering is to build a dendrogram to get a tree of relations between elements.

Some methods are based on densities, the most common method being the DBSCAN [29] (Density-Based Spatial Clustering of Applications with Noise). For this algorithm, a cluster is formed when a set of points are packed together. This algorithm can be efficient when we expect different clusters in different part of the state space. It does not assume a particular shape for the clusters and can work with large data sets.

For some applications, one clustering method is not able to accurately group the different parts of the set: In this case, a prediction can be built by combining the strength of different clustering methods. This type of method is called ensemble method [30, 31]. Without entering in the details, this method is based on the mean of the similarity matrix performed for each clustering method, to deduce a consensus matrix. It should be mentioned that these algorithms can result in a greater number of clusters than K, especially when the clustering methods give quite different results.

Each of the clustering algorithms has somewhat different strength, k-means allows for a characterization of the clusters using the cluster means. It can also be viewed as a decomposition method, where each data point is represented by its cluster center. DBSCAN allows for the detection of "noise points" that are not assigned any cluster, and it can help automatically determine the number of clusters. In contrast to the other two methods, it allows for complex cluster shapes. DBSCAN sometimes produces cluster of very differing size, which can be strength or a weakness. Hierarchical algorithms can provide a whole hierarchy of possible partitions of the data, which can be easily inspected.

2.3.4 Clustering in Practice for Telecom Data

Some requirements are necessary to have a clustering algorithm performing well in practice with telecom data. In all the following paragraphs, exploration of data is helpful to improve our understanding about it and to deduce which prior assumptions are plausible.

A first requirement is to carefully select the features of interest. They must be selected to represent the underlying patterns we want to discover. By selecting too many features, the curse of dimensionality can lead to poor clustering. This selection is mainly done from expertise and by exploration of data.

A second requirement is to perform feature scaling of data to remove preference for a feature compared to another. Clustering without feature scaling leads to compare features on different scale and many clustering algorithms may then fail (especially with K-means based algorithms). The feature scaling can be done automatically by normalizing all features between 0 and 1. Sometimes, log-transformation can be computed to scale specific features (see [32, 33] for the generalization of this technique known as Box-Cox transformation).

A third requirement is the selection of the number K of clusters. Although methods to automatically select a value of K do exist (a comparison between different algorithms are presented in [28]), the selection is usually chosen manually by researchers and refined after performing different computations for different values.

2.4   Forecasting Methods

Collection of telecom indicators is most often performed regularly in time and represented mathematically as a sequence indexed by time (for example representing the evolution of an indicator such as a telecom traffic KPI). Such sequence of objects is called a time series, and forecasting methods are developed to analyze behavior and trends of the time series in order to show consumer usage patterns and to make predictions of the future values of the time series. The main aim is to obtain accurate forecasting predictions, containing both punctual prediction and confidence interval around this prediction.

In this context, methods developed in the previous sections such as regressions can be used but are often insufficient to automatically and efficiently manage auto-regressive behavior of most time-series (which means that the value at a time t is strongly correlated with the value at the time $t - 1$ and at the previous times). To address this problem, forecasting methods have been developed by placing the time parameter as a main sequential parameter [2, 34].

In the following, we give a broad view of univariate evenly spaced time series, which means that the spacing of observation times is constant (for multiple time series analysis, we refer in [35], containing for example the VAR model). We assume that no missing values are present. When those conditions are not fulfilled, we usually use interpolation methods to complete the time series [36]. We first describe how a time series can be decomposed into different components (Sect. 2.4.1), and therefore introduce two different types of forecasting algorithms (exponential smoothing in Sect. 2.4.2 and ARIMA models in Sect. 2.4.3).

2.4.1   Decomposition of a Time Series

In order to understand a time series, we usually perform a decomposition which extracts the most common patterns and helps to discover the underlying evolution of the time series. Given (x_t) a time-series indexed by time, we wish to decompose it as follows for all time t:

$$x_t = T_t + S_t + \varepsilon_t,$$

where:

- (T_t) is the trend component, which corresponds to the long-term evolution of the series. In most decomposition, this term does not have to be linear and includes cyclic behaviors,
- (S_t) is the seasonal component, which reflects seasonality of the time series (for example peak and off-peak months for a monthly time series). The periodicity of this component is initially known and fixed, however most decomposition methods can deduce a seasonal function which is not strictly a periodic function,

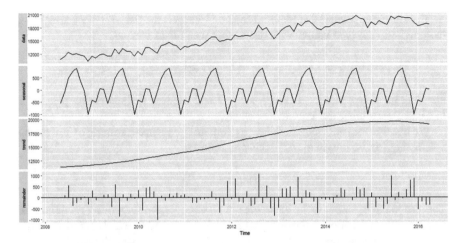

Fig. 2.7 STL decomposition of a time series. The original time series is shown on the top (*data*). The three other graphs correspond to the decomposition by representing the *seasonal*, *trend* and *remainder*. The seasonal component helps to show peak and off-peak months; The trend component highlights the growth of the time-series with an halt in the most recent collected data; The remainder component is quite regular and does not show any strong outlier

– (ε_t) is the remainder of the time series after other components have been removed. The underlying distribution of this remainder can be modeled through auto-regressive models and studied separately.

One of the most robust methods to deduce the different components is the STL algorithm [37] (Seasonal and Trend decomposition using LOESS). This algorithm iteratively performs local regressions to fit both seasonal and trending components. The smoothness of those two components can be selected through hyper-parameters. On Fig. 2.7, an example of such decomposition is computed on a monthly time series (with a selected periodicity of 12, since we expect an annual periodicity). This decomposition highlights a halt in growth of the curve (in the trend component), a peak period during April/May of each year (in the seasonal component) and the absence of strong outliers (in the remainder component).

2.4.2 Exponential Smoothing Models

In this section, we first introduce a forecasting method known as simple exponential smoothing [38]. This method is one of the easiest methods to obtain predictions of future values. It is barely used in practice (because for example it cannot manage the trend and the seasonal component) but the underlying method can be generalized to deduce more complex patterns.

The simple exponential smoothing method forecasts the time $T + 1$ by weighting the previous observed values with decreasing weights as follows:

$$\widehat{x}_{T+1} = \alpha x_T + \alpha(1 - \alpha)x_{T-1} + \alpha(1 - \alpha)^2 x_{T-2} + \alpha(1 - \alpha)^3 x_{T-3} + \dots$$

knowing the time series (x_t) from time 1 to T and where α is a smoothing parameter in $(0, 1)$ representing the exponential decay.

We observe that the weights follow a geometric sequence summing to 1, then the forecasting can be rewritten recursively as follows:

$$\widehat{x}_{T+1} = \alpha y_T + (1 - \alpha)\widehat{x}_T.$$

This recursive equation is the basic idea of all exponential smoothing models. One common generalization is the additive Holt-Winters model, which is based on the simple exponential smoothing but also incorporates a recursive description of the seasonal and trending components. The equations to build this model are technical and are not detailed here. In the Fig. 2.8 (a), an example of prediction obtained with additive Holt-Winters is shown for 12 months, capturing both the annual seasonality and the halt of growing of the curve. We may note that Holt-Winters model can also be generalized into the so-called ETS (Error, Trend, Seasonal) models (see in [2, 34, 39]).

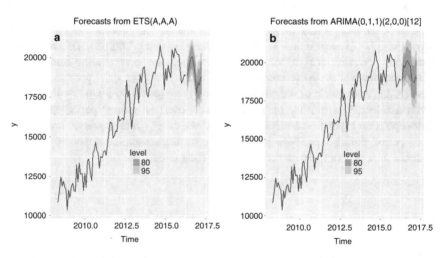

Fig. 2.8 Forecasting for the next 12 months of a time series with additive Holt-Winters and ARIMA, computed from *forecast* package of R. On each graph, the original time-series (in black), punctual predictions and confidence of intervals (at levels 80% and 95%) are shown. **(a)** Prediction with additive Holt-Winters. **(b)** Prediction with a seasonal ARIMA model with *auto. arima*; the found non seasonal parameters are $(0, 1, 1)$ and the seasonal parameters are $(2, 0, 0)$

2.4.3 ARIMA Models

ARIMA models [40] (standing for autoregressive integrated moving average) are another way to model the time series by describing the autocorrelations in the data. A first step is to deduce a *stationary* time series. The formal definition involves that, for all t and s, the underlying joint distribution of the random variables $(x_t, \ldots, x_{t+s})$ does not depend on t. In practice, a trend or a seasonal component in the time series induces a non-stationary behavior, and we must differentiate it (one or several times) to obtain a stationary time series. Here differentiation means to compute the differences:

$$x'_t = x_t - x_{t-1}.$$

The resulting stationary time series (y_t) is then modeled as follows into two parts:

$$y_t = c + \varphi_1 y_{t-1} + \ldots + \varphi_p y_{t-p} + \theta_1 \varepsilon_{t-1} + \ldots + \theta_q \varepsilon_{t-q} + \varepsilon_t,$$

Where:

- c is a real parameter,
- $(y_{t-1}, \ldots, y_{t-p})$ are the p previous values of the time series, related with parameters $(\varphi_1, \ldots, \varphi_p)$,
- $(\varepsilon_t)_t$ is a time series computed independently and identically distributed according to a Gaussian distribution, and here $(\varepsilon_{t-1}, \ldots, \varepsilon_{t-q})$ are related with parameters $(\theta_1, \ldots, \theta_q)$,
- ε_t is a random noise added in addition with the previous noises.

Through this description, we can observe that: $(\varphi_1, \ldots, \varphi_p)$ corresponds to the autoregressive part of the model (AR), i.e. a regression of the variable (y_t) against itself; and that $(\theta_1, \ldots, \theta_q)$ corresponds to the mobile average part of the model (MA), i.e. a linear weighted combination of past errors.

On the whole, three main parameters have to be found to deduce the ARIMA model:

- p the order for the autoregressive part,
- d the number of times we integrated the time series,
- q the order for the mobile average part.

Some algorithms have been developed to efficiently select both this combination (p, d, q) and the parameters needed for the model $(c, \varphi_1, \ldots, \varphi_p, \theta_1, \ldots, \theta_q)$.

To improve the accuracy of forecasting, the seasonal part of the time series is often modeled at the same time with another ARIMA process leading to new parameters (P, D, Q). In Fig. 2.8 (b), we observe forecasted results on a time series with a seasonal-ARIMA process. The parameters have been automatically selected with the *auto.arima* function of R. Finally, the selected parameters are $(0, 1, 1)$ for the main part and $(2, 0, 0)$ for the seasonal part.

2.5   Neural Network and Deep Learning

2.5.1   Neural Network

Neural networks are defined as algorithms to recognize patterns from data imitating the human brain. The patterns neural networks can recognize should be numerical in vectors, and data can be images, sound, text or time series [41, 42].

Connected artificial neurons (like biological neurons in an animal brain) are the base units in the Neural Network. The connection between different artificial neurons (like the synapse) transmits signal from one to another. The artificial neuron receiving signals processes the signal and transmits the processed result to the neurons it connects with. In the mathematical model of neural networks, signals transmitted between different artificial neurons are numerical numbers, and the output of each artificial neuron is processed and calculated by using a non-linear function which sums the inputs values it receives. In Neural Networks, each connection has a weight (or coefficient in its mathematical model) which can be adjusted in the learning process. Such weights either increase or decrease the signal strength for a connection. Artificial neurons can set a threshold to filter out the signals which has higher or lower numerical value compared with the threshold. Artificial neurons in a Neural Network are organized in layers. Neurons in different layers execute different calculations of transformation. Neural networks can be utilized for different classification tasks in machine learning. After training with historical data, Neural Network can be implemented for computer vision, speech recognition and machine translation. Advanced neural network models, like deep neural network can also be leveraged for playing board and video games and medical diagnosis.

Neurons in neural network receive input, execute calculation with the activation function, and produce output based upon the input and calculated result. The neural network is a weighted directed graph, where neurons are connected with a weighted value. The weights can be adjusted in the learning process with a defined learning rule. Therefore, neurons, connections with weights, activation function, and the learning rule are the primary components need to be defined for any neural network.

In a neural network, a neuron, j, which receives input $p_j(t)$ from its predecessor neurons and get its activation $a_j(t)$ by using activation function f which computes the activation at t + 1 from $a_j(t)$, θ_j and the input $p_j(t)$

$$a_j(t+1) = f\left(a_j(t), p_j(t), \theta_j\right)$$

and an output function f_{out}:

$$o_j(t) = fout\left(a_j(t)\right)$$

Neurons in the input layer have no predecessor; they serve as the input interface for the whole network. Neurons in the out layer have no successor, and serve as the output of the network. Output of a neuron can be transferred as input for other neurons. The propagation function in the neural network calculates the input $p_j(t)$ to the neuron j from the outputs $o_i(t)$ of predecessor neurons with the weight between them. The learning rule defined in a neural network adjusts the parameters of the neural network. The learning process is to modify the weights and threshold to minimize the difference between predicted value and the real value.

2.5.2 Deep Learning

Deep learning is a class of artificial neural networks based machine learning techniques. It utilizes multiple layers of neural network with different architectures to get richer functionally as Fig. 2.9 shows. Deep neural networks, deep belief networks, recurrent neural networks, and convolutional neural network are the most popular deep learning algorithms which have been successfully applied to many fields like, computer vision, speech recognition, natural language processing, audio recognition, bioinformatics and drug design. Different from traditional machine learning techniques, deep learning can automatically learn patterns or characteristics from data such as images, video or text, without human domain knowledge, they can directly learn from raw data and can increase their predictive accuracy. DNN, CNN, and RNN are the most representative deep learning algorithms and will be described below in more details.

Multiple hidden layers with the input and output layers constitute the deep neural network (DNN) [43]. DNN is a feedforward neural network, in such a network signals flow from the input layer to the output layer via multiple hidden layers without looping back. With multiple hidden layers, such model could have richer

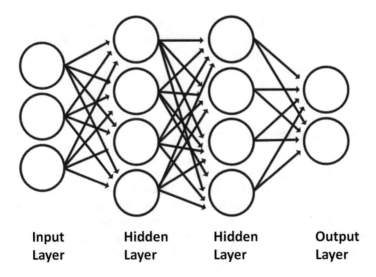

| Input | Hidden | Hidden | Output |
| Layer | Layer | Layer | Layer |

Fig. 2.9   DNN Deep learning algorithm

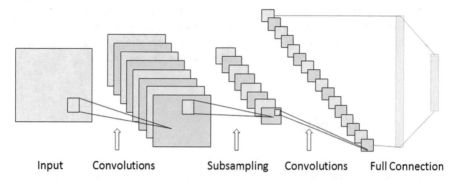

Fig. 2.10 CNN Deep learning algorithm

functionality and achieve higher accuracy in some tasks. However, these hidden layers also significantly increase the computation complexity.

In a recurrent neural network (RNN) [44], the network topology is a directed graph where neurons are connected along a time sequence. It uses internal state to process sequences of input and memorizes the temporal behavior for a time sequence. This feature makes RNN applicable to tasks such as speech or handwriting recognition, and sentence translation. When RNN calculates a parameter, it not only takes consideration of the current input but also the historical results it has learned from the previous input. Therefore, there are two inputs for RNN: input from now and input from historical calculation. Taking consideration of the historical status is the core mechanism of RNN. This is because the sequence of data contains very import information about what is coming later.

Multiple convolutional layers, polling layers and fully connected layers constitute a Convolutional Neural Network (CNN) [45] network shown in Fig 2.10. It is a special case of the neural network described in the previous section. With the 2D structure of the input data (for example an image), it is easy to train a CNN model since it has very fewer parameters than fully connected networks with the same number of hidden units.

2.6   Reinforcement Learning

Reinforcement Learning (RL) is a concept of machine learning techniques where an agent explores the environment, performs actions, updates its status, and finally knows the environment by using the results feed backed from the environment. The main task in Reinforcement learning [46] is getting an agent to act in the environment so as to maximize its rewards returned by the environment as Fig. 2.11 shows. In the next time step, agent in RL receives a delayed reward from the environment to evaluate its action executed in the current step.

Fig. 2.11 Reinforcement
learning

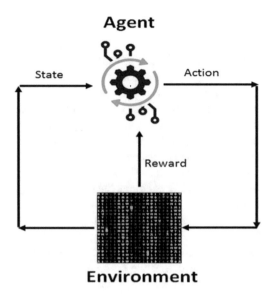

RL algorithm has two primary components: an agent and an environment. The environment can be considered as the platform or interface of the object that the agent is acting on and interacting with. The agent represents the algorithm itself. The environment determines the state of the agent and how to reward or punish the action the agent make. In different states, the agent takes different action in response to the environment. Through intersection with the environment, the agent updates its knowledge with the feedback returned by the environment to evaluate its last action.

2.6.1   Model and Policy

To get the feedback from the environment and determine how to perform actions for each state, the RL algorithm should define the model and policy. The algorithm learns the probability of transition $T(s_1|(s_0, a))$ from the current state s_0 with an action a to the next state s_1. The agent will know how to enter a specific state at current state executing a specific action, if the probabilities are successfully learned before. Hence, the model-based algorithms need big space to save the state and their actions.

Model-free algorithms are different from the model-based ones, since they rely on trial-and-error to update their knowledge of environment. Therefore, it does not require big space to store all the information about their states and actions. In RL algorithm, policy will determine how to execute action according the state. Agent learns the values based on its current action derived from the policy, whereas its off-policy counterpart learns it based on the action obtained from another policy.

2.6.2 RL Algorithms

Q-Learning [47] is categorized as a model-free reinforcement learning algorithm. Agents in Q-Learning learn the environment. They act in Markovian domains by experiencing the consequences of actions. It is not necessary to build maps of the domain. Even the actions are determined according to a more exploratory or randomly case, Q-Learning still can learn the optimal policy. In Q-learning algorithms, an agent learns the policy based on its historical interaction with the environment and keeps updating the Q-table which records the historical interaction with the environment. These historical interactions are a sequence of state-action-rewards value. For example, the sequence of $(s_0,a_0 \rightarrow r_1,s_1,a_1 \rightarrow r_2,s_2,a_2 \rightarrow r_3,s_3,a_3 \rightarrow r_4,s_4...)$, shows that the agent was in state s_0 and did action a_0. The environment evaluates the action and rewards r_1. The state is changed from s_0 to s_1 accordingly. The agent did action a_1 at state s_1, received reward r_2, and changed to state s_2; and so on. In this algorithm, the interaction history is considered as a sequence of experiences interacted with the environment. The goal for an agent in Q-Learning is to update and maximize the Q value. $Q^*(s,a)$, where a is an action and s is a state. It is the expected value of doing an action a in state s and then following the optimal policy.

$$Q^*_{(s,a)} = \sum_{s' \in S}^{n} p_r(s'|s,a)[R(s,a,s') + \gamma V^*)s'_-]$$

Another RL algorithm is called as State–action–reward–state–action (Sarsa), which is an algorithm for learning a Markov decision process policy [48]. In the Sarsa, agent performs action a_1 at state s_1, get reward r_1. After that, it goes to state s_2, and performs action a_2, and gets rewards r_2 before it goes back and updates the value of a_1 it performed in state s_1. Compared with the Q-learning algorithm where the future reward is defined as the highest possible action that can be taken from state s_2, Sarsa takes the value of the actual action that was taken. In each RL algorithm, the Q value for a pair of state-action can be updated and adjusted. Q values represent in the reinforcement learning are the rewards received by agent.

References

1. Hastie T, Tibshirani R, Friedman J. The elements of statistical learning. 2001. NY Springer, 2001.
2. Hyndman, R.J. and Athanasopoulos, G. (2013) Forecasting: principles and practice. OTexts: Melbourne, Australia. http://otexts.org/fpp/. Accessed on 2016/09/20
3. Stanton J M. Galton, Pearson, and the peas: A brief history of linear regression for statistics instructors[J]. Journal of Statistics Education, 2001, 9(3).
4. Cleveland W S. Robust locally weighted regression and smoothing scatterplots[J]. Journal of the American statistical association, 1979, 74(368): 829–836.
5. Hastie T J, Tibshirani R J. Generalized additive models[M]. CRC Press, 1990.

6. Tibshirani R. Regression shrinkage and selection via the lasso[J]. Journal of the Royal Statistical Society. Series B (Methodological), 1996: 267–288.
7. Freedman D A. Statistical models: theory and practice[M]. cambridge university press, 2009.
8. Bengio, Yoshua, Ian J. Goodfellow, and Aaron Courville. "Deep learning." An MIT Press book. (2015).
9. Simon N Wood. Modelling and smoothing parameter estimation with multiple quadratic penalties. Journal of the Royal Statistical Society. Series B, Statistical Methodology, pages 413–428, 2000.
10. Hoerl A E, Kennard R W. Ridge regression: Biased estimation for nonorthogonal problems [J]. Technometrics, 1970, 12(1): 55–67.
11. Zou H, Hastie T. Regularization and variable selection via the elastic net[J]. Journal of the Royal Statistical Society: Series B (Statistical Methodology), 2005, 67(2): 301–320.
12. Alan Miller. Subset selection in regression. CRC Press, 2002.
13. Saeys, Yvan, Iñaki Inza, and Pedro Larrañaga. "A review of feature selection techniques in bioinformatics." bioinformatics 23.19 (2007): 2507–2517.
14. Hotelling, H. (1933). Analysis of a complex of statistical variables into principal components. Journal of Educational Psychology, 24, 417–441, and 498–520.
15. Walker S H, Duncan D B. Estimation of the probability of an event as a function of several independent variables[J]. Biometrika, 1967, 54(1–2): 167–179.
16. Cox D R. The regression analysis of binary sequences[J]. Journal of the Royal Statistical Society. Series B (Methodological), 1958: 215–242.
17. Hosmer Jr, David W., and Stanley Lemeshow. Applied logistic regression. John Wiley & Sons, 2004.
18. Altman N S. An introduction to kernel and nearest-neighbor nonparametric regression[J]. The American Statistician, 1992, 46(3): 175–185.
19. Suykens J A K, Vandewalle J. Least squares support vector machine classifiers. Neural processing letters, 1999, 9(3): 293–300.
20. Rokach L, Maimon O. Data mining with decision trees: theory and applications[M]. World scientific, 2014.
21. Liaw A, Wiener M. Classification and regression by randomForest. R news, 2002, 2(3): 18–22.
22. Ho, Tin Kam (1995). Random Decision Forests (PDF). Proceedings of the 3rd International Conference on Document Analysis and Recognition, Montreal, QC, 14–16 August 1995. pp. 278–282.
23. James MacQueen et al. Some methods for classification and analysis of multivariate observations. Proceedings of the fifth Berkeley symposium on mathematical statistics and probability, volume 1, pages 281–297. Oakland, CA, USA, 1967.
24. Geoffrey J McLachlan and Kaye E Basford. Mixture models: Inference and applications to clustering. Applied Statistics, 1988.
25. Dempster A P, Laird N M, Rubin D B. Maximum likelihood from incomplete data via the EM algorithm[J]. Journal of the royal statistical society. Series B (methodological), 1977: 1–38.
26. Leonard Kaufman and Peter Rousseeuw. Clustering by means of medoids. North-Holland, 1987.
27. James C Bezdek. Pattern recognition with fuzzy objective function algorithms. Springer Science & Business Media, 2013.
28. Leonard Kaufman and Peter J Rousseeuw. Finding groups in data: an introduction to cluster analysis, volume 344, 6C1003. John Wiley & Sons, 2009.
29. Ester M, Kriegel H P, Sander J, et al. A density-based algorithm for discovering clusters in large spatial databases with noise[C]. 1996, 96(34): 226–231.
30. Ana Fred. Finding consistent clusters in data partitions. Multiple classifier systems, pages 309–318. Springer, 2001.
31. Dietterich T G. Ensemble methods in machine learning. Multiple classifier systems. Springer Berlin Heidelberg, 2000: 1–15.

32. Sakia, R. M. "The Box-Cox transformation technique: a review." The statistician (1992): 169–178.
33. Guerrero, Victor M., and Richard A. Johnson. "Use of the Box-Cox transformation with binary response models." Biometrika 69.2 (1982): 309–314
34. Hamilton J D. Time series analysis[M]. Princeton: Princeton university press, 1994.
35. Lütkepohl H. New introduction to multiple time series analysis[M]. Springer Science & Business Media, 2005.
36. Royston P. Multiple imputation of missing values[J]. Stata journal, 2004, 4(3): 227–41.
37. Cleveland R B, Cleveland W S, McRae J E, et al. STL: A seasonal-trend decomposition procedure based on loess[J]. Journal of Official Statistics, 1990, 6(1): 3–73.
38. Holt Charles C. Forecasting trends and seasonal by exponentially weighted averages [J]. International Journal of Forecasting, 1957, 20(1): 5–10.
39. Gardner E S. Exponential smoothing: The state of the art[J]. Journal of forecasting, 1985, 4(1): 1–28.
40. Brockwell P J, Davis R A. Introduction to time series and forecasting[M]. Springer Science & Business Media, 2006.
41. "Artificial Neural Networks as Models of Neural Information Processing | Frontiers Research Topic". Retrieved 2018-02-20.
42. McCulloch, Warren; Walter Pitts (1943). "A Logical Calculus of Ideas Immanent in Nervous Activity". Bulletin of Mathematical Biophysics. 5 (4): 115–133.
43. Dahl, G.; et al. (2013). "Improving DNNs for LVCSR using rectified linear units and dropout" (PDF). ICASSP.
44. Schmidhuber, J. (2015). "Deep Learning in Neural Networks: An Overview". Neural Networks. 61: 85–117.
45. Deng, L.; Li, J.; Huang, J. T.; Yao, K.; Yu, D.; Seide, F.; Seltzer, M.; Zweig, G.; He, X. (May 2013). "Recent advances in deep learning for speech research at Microsoft". 2013 I.E. International Conference on Acoustics, Speech and Signal Processing.
46. van Otterlo, M.; Wiering, M. (2012). "Reinforcement learning and markov decision processes". Reinforcement Learning. Springer Berlin Heidelberg: 3–42
47. Barto, A. (24 February 1997). "Reinforcement learning". In Omidvar, Omid; Elliott, David L. Neural Systems for Control. Elsevier.
48. Wiering, Marco; Schmidhuber, Jürgen (1998-10-01). "Fast Online Q(λ)". Machine Learning. 33 (1): 105–115.

Chapter 3
Trending LTE Network Performance

LTE network performance is commonly evaluated from Key Principal Indicators (KPIs), which are relevant numeric indicators summarizing the global performance of the network. A hot topic is to evaluate and forecast those network performance indicators from the available traffic flow and history. This evaluation of LTE network performance is crucial for mobile carriers to leverage appropriate strategies of capacity requirement and management.

In this chapter, we focus on a systematic methodology of data analytics and modeling to evaluate LTE network capacity based upon traffic measurement and service trends. Two main tools presented in Chap. 2 are necessary to develop data analytics methodology: Forecasting algorithms and regression algorithms. The forecasting algorithms allow understanding the whole evolution of a fixed indicator to predict the next possible values. There can take into account the important components of the time series such as the trend, the seasonality and the possible arising bursts. Legacy methods have been applied to telecom industry only with forecasting tools, but due to the uniqueness of telecom network, those traditional methods are usually unable to fit very well for business cases. Consequently, customized methods have been developed.

Those new methods exploit the whole source of available data to forecast the feature of interest. Indeed, simple indicators such as low level traffic measurement and specific services traffic are used. Those indicators show how network resource is quantitatively consumed by the various applications or users whose traffic consumption is reflected by the network KPIs. Therefore, they are combined and relationships with a main LTE KPI indicator are provided through regression algorithms. The methodology developed in this chapter combines those two tools to provide a complete forecasting methodology. In the final section, this methodology is validated and shows high accuracy, robustness and reliability.

© Springer International Publishing AG, part of Springer Nature 2018   39
Y. Ouyang et al., *Mining Over Air: Wireless Communication Networks Analytics*,
https://doi.org/10.1007/978-3-319-92312-3_3

3.1   Network Performance Forecasting Strategy

3.1.1   Strategies with Direct Forecasting

Many researchers studied wireless network performance, service quality, and capacity management in both academia and industry [1, 2, 4–7]. There also are many published research methods to forecast a given network resource utilizing time series or other trending algorithms [3]. However, the causality of this time series is unknown to network capacity in these methods. They cannot analyze how the network resource is quantitatively consumed by various applications or users whose traffic consumption is reflected by those network KPIs. Moreover, the user and service behaviors can be lost using these methods. Specifically, many components such as user behaviors to consume traffic, diversity of traffic consumption between services, seasonality of traffic consumption, and random spikes cannot be decomposed and derived.

3.1.2   Analytical Model

A relational model is proposed in this chapter to overcome the shortcomings mentioned in the above section. This proposed model is able to derive the quantitative relation between LTE network KPIs and network resource indicators, and forecasts network KPIs as shown in Fig. 3.1. The forecasted LTE KPIs can be obtained by inputting the predicated network resource indicators in the relational model for KPI and resource indicators.

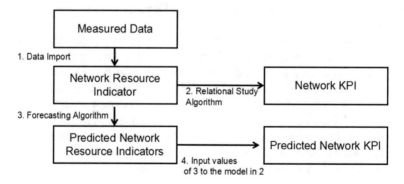

Fig. 3.1 Work flow of the analytical model

3.2 Relationship Between Network Resources and Performance Indicators

3.2.1 Relation Between LTE Network KPI and Resources

A given LTE network resource is consumed by either voice or data traffic generated by mobile device user. The traffic can be quantitatively represented by LTE network traffic indicators. Hence, we focus on a qualitative schematic relation between a given LTE network KPI and its associated network resources. As Fig. 3.2 shows, the X axis denotes the LTE network resource that is being consumed and the Y axis indicates the measured KPI. The LTE KPI as a function of LTE network consumed resources is split into four areas which are separated by three dotted lines in the figure. We assume that if KPI value is larger the KPI is better, which means that a high value of Radio Resource Control (RRC) Connections Setup Success Rate denotes a good RRC connection success rate.

In Fig. 3.2 Zone 1 (the left-most part in the figure), the curve shows a constant good KPI. The value is first equivalent to the reference point at very beginning where almost no network resource is consumed. It means the LTE network is extremely light loaded. A slight growth of network resource consumption may not affect the KPI at all in Zone 1. Zone 2 has a sinking KPI. At a particular point or within a certain range, the formal quasi-perfect KPI scenario cannot be maintained any more. As the LTE network resources continue to be consumed, the KPI starts to degrade at a limited range where the variance of KPI is larger than that in the Zone 1 since the

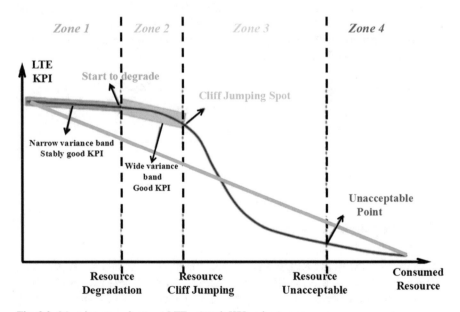

Fig. 3.2 Mapping curve between LTE network KPI and resource

Quality of Service (QoS) in the Zone 2 is slightly deteriorated due to fewer remaining resources which are available. Zone 3 has a cliff jumping KPI. As the LTE network KPI continues to degrade and once it exceeds a particular threshold, such a quasi-flat relation between KPI and consumed resource starts to collapse. As fewer and fewer available network resources are left, a more additional resource to be consumed imposes a considerable impact on the LTE KPI. The KPI in Zone 4 is not acceptable. As soon as the KPI drops at another particular point or a certain range, the KPI value behind that point or range is no longer acceptable. The network may be out of service since a minimal QoS cannot be maintained by it.

Considering the piecewise relation from Zone 1 to Zone 4, we make an assumption that the change of LTE network KPI depends on the current level of KPI, given a certain amount of resource to be changed. After a transformation, the relation between KPI and network resource can be derived by Eq. 3.1.

$$\frac{\partial KPI}{\partial Resource} \sim \alpha.KPI + \beta, \quad \frac{\partial KPI}{\alpha.KPI + \beta} \sim \partial Resource$$

$$\int \frac{\partial KPI}{\alpha.KPI + \beta} .dKPI \sim \int \partial Resource.dResource \tag{3.1}$$

$$KPI = A.e^{-B.Resource} + C$$

The function of n influence factors in terms of resource 1 to n is as follows:

$$KPI = f(Resource1, Resource2, \ldots Resource\ n)$$

Hence, we expand the pair-value relation between a given KPI and a given network resource, to a KPI and multiple related network resource, each of which has a coefficient do denote its weight to KPI. So the KPI equation can be represented as an exponential model as Eq. 3.2:

$$KPI = Coeff_1 \times (A_1.e^{-B1.Resource1} + C_1) + Coeff_2 \times (A_2.e^{-B2.Resource2} + C_2)$$

$$+ \ldots + Coeff_n \times (A_n.e^{-Bn.Resourcen} + C_n)$$

$$= \sum_{i=1}^{n} Coeff_i \times (A_i.e^{-Bi.Resourcei} + C_i)$$

$$\tag{3.2}$$

3.2.2 Regression Model

The relation between the KPI and the resource indicators can be derived by using Generalized Additive Model (GAM) and sigmoid model. These two models may overcome two drawbacks of leveraging exponential model in our equations. First, in Zone 1 the KPI presents a smooth relation to resource indicators while in Zone 2 and

3 the KPI presents more like an exponential relation to resource indicators. This is because in Zone 1 the change of KPI is not sensitive to the change of resource indicators when the remaining resource is still highly sufficient. So the KPI maintains at a relatively optimal and stable level when physical resource is consumed at a low level represented by small values of resource indicators in Zone 1. As a result, the smooth gradient of KPI in Zone 1 is difficult to be represented by exponential function. Conversely, Zone 3 and Zone 4, which are concave up, are more fitted by exponential model. Second, there is a cliff jumping spot between Zone 2 and Zone 3 to represent that the KPI starts to degrade sharply once the resources are consumed to a particular point. Since, the exponential model behaves as a monotony function without a turning point, there isn't any inflection point to represent the cliff jumping point. Usually, linear model are desirable because they are simple to fit, results are easy to understand, and there is a wide variety of useful techniques for testing the assumptions involved. Nonetheless, there are cases where the linear model should not be applied because of an intrinsic nonlinearity in the data. GAM provides a means for modeling such data. It models the nonlinear relationship, by entirely relaxing the linearity assumption. The advantage of this scheme is obvious that both the smooth relation in Zone 1 and 2 and the sharply decreasing trend in Zone 3 and 4 can be locally fitted.

Another merit is that GAM can be easily extended to multi-dimension regression, which means the multiple resource indicators can simultaneously join the model for regression. There is no separated step 2 in Eq. 3.2 to train the coefficient per resource indicator. The GAM based algorithm for our model is given in below equation:

$$KPI = \alpha + f_{s\_1}(Resource\_1) + f_{s\_2}(Resource\_2) + \cdots + f_{s\_n}(Resource\_n) + \varepsilon \tag{3.3}$$

where $f_{s\text{-}i}$ means the smoothing function for i-th resource indicator. It represents the arbitrary function that can be estimated by GAM. α indicates an intercept and ε represents a random error.

The advantage of sigmoid model is that the cliff jumping spot can be derived by letting derivative $= 0$, in which the point at 2nd derivative changes sign. In addition, the smooth area in Zone 1 and 2 can also be fitted by sigmoid curve. The standard sigmoid function is represented by Eq. 3.4:

$$y = f(x) = \frac{1}{1 + e^{-x}} \tag{3.4}$$

However, the Sigmoid curve of the cliff jumping model may not be as perfect as the standard format. So, let the Sigmoid-curve function be:

$$KPI = f(Resource) = \frac{A_1}{1 + e^{-B(-Resource+C)}} = \frac{A}{1 + e^{B(Resource-C)}} \tag{3.5}$$

where B denotes the curvature of the Sigmoid curve, C denotes inflection point of the Sigmoid curve. According to Eqs. 3.4 and 3.5, the equation of 3.2 can be revised as:

$$KPI = Coeff_1 \times \left(\frac{A_1}{1 + e^{B_1(Resource\_1 - C_1)}} \right) + Coeff_2 \times \left(\frac{A_2}{1 + e^{B_2(Resource\_2 - C_2)}} \right)$$

$$+ \cdots + Coeff_n \times \left(\frac{A_n}{1 + e^{B_n(Resource\_n - C_n)}} \right) = \sum_{i=1}^{n} Coeff_i \times \left(\frac{A_i}{1 + e^{B_i(Resource\_i - C_i)}} \right)$$

3.3 Forecasting Network Resources

3.3.1 Model to Forecast LTE Network Traffic and Resource

After studying the relation between LTE KPIs and network capacity, we now focus on forecasting network traffic. In a given LTE network, any network traffic related with indicators or resources indicators can be measured at level of carrier, sector, or cell. The time granularity for KPI is normally by hour. For a given cell in a LTE network, a given network traffic indicator or resource indicator in a particular period (t_i, t_j) can be presented by $X(t) = \{X(ti, X(ti + 1), \ldots, X(tj)\}$. We could decompose $X(t)$ into four components: Trend Component T(t), Seasonality component S(t), Burst component B(t) and Random error R(t). Eventually, the network traffic indicator could be represented by $X(t) = (1 + B(t))*(T(t) + S(t) + R(t))$, which is depicted in Fig. 3.3.

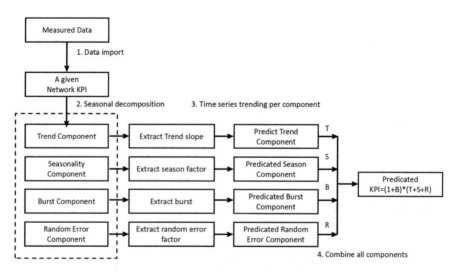

Fig. 3.3 Logic flow to forecast LTE network traffic and resource

3.3.2 Forecasting Network Resources

The trend component reflects how user behaviors, rate plan policy, and change of user numbers impact LTE network traffic and network resource consumption in a long term. It represents the base value of $X(t)$ in a certain lengthy period (e.g. 30–90 days) with a large granularity. It splits the measured time series of the given $X(t)$ into several pieces, the length of each is given by m. In a short term prediction (e.g. 0–60 days), m is given by 10, in a middle term prediction more than 180 days, m is given by 30 or longer. Afterwards, piecewise fit the m pieces of data. The trend component is created by:

$$T(t) = f\{X_K, Slope_K\} \tag{3.7}$$

where X_k denotes the first fitted value in each piece K. $Slope_k$ denotes the slope of piece K. In the fitting process, we first divide the time series of $X(t)$ into m pieces. For piece $K = 1$, fit a line with Slope k and starting point $= X_k$. For piece where $K = 2$ to m, fit a line with $Slope_k$ and starting point of line K equals the last fitted point of line K-1, this step makes sure each fitted line is interconnected without any gap between lines. The final step in trend component is to minimize the mean error rate between fitted value and true value in each piece.

In some particular cases, the network traffic value may change sharply before a turning point, then change slowly after the turning point. For example, in a school area, the network traffic value may change sharply at the beginning and end of school vacation(s) as shown in Fig. 3.4. During school vacation periods and periods when school is in session, the network traffic values may remain relatively stable. Such sharp changes may not be recognized as bursts or seasonal changes, depending on the period of the seasonality component or the granularity of the burst component. Instead, it should be explained by trend component as a change in the long term. However, in 2nd order exponential smoothing, the rapid decreasing intervals play much more weight than slowly increased intervals in determining the perspective

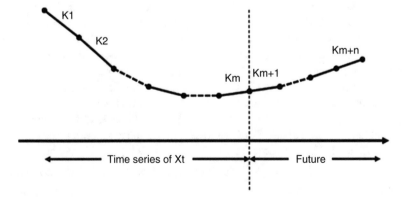

Fig. 3.4 Predict trend component

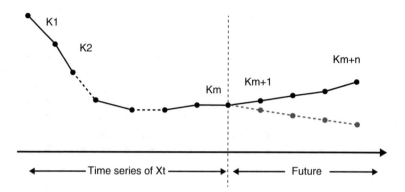

Fig. 3.5 Compensate super decreased slopes

slopes. Hence, no matter how the smoothing coefficient is optimized, a possible fact is the perspective slope will be negative forever. This causes the forecasted values to be negative values, which is meaningless as shown in Fig. 3.5. As a result, there is a need to improve the algorithm above by compensating those super decreased or increased gradient in a short period. A compensation formula is given that if recent N successive slope is no less than zero, the perspective slope should not be less than zero. The equation is given by:

$$K'_{T+i} = Max\{K_{T+i}, \gamma.\min\{K_T, K_{T-1}, K_{T-2}, \cdots, K_{T-N+1}\}\} \qquad (3.8)$$

Where γ is a constant that can be tuned until optimal, K'_{T+i} denotes the improved slope after compensation.

The seasonality component represents periodic changes in traffic during a period of a predetermined length (e.g., a weekly period) as network traffic over a weekly period is correlated due to people engaging in similar activities on similar days of the week. For example, the wireless network traffic in Chicago on Saturday, March 1st, 2014, at 8 PM may be highly similar to the wireless network traffic in Chicago on Saturday, March 8th, 2014, at 8 PM. Alternatively, other periods different from 7 days may be used. For example, in a country with a six-day cycle of work or school days, a period of 6 days may be used. In some case, the cycle of the periodicity component may be different from the cycle of work or school days. To generate the seasonality equation, the length of a period may be determined.

A time series of network traffic values of length L may be provided. The possible period lengths, i, may be between 1 and L/2, where in each i, there are j pieces. For example, L may be 70 days, i may be between 1 and 35 days, and the j pieces may represent the 24 h in each day. For $i = 1$ to L/2, the variance for each piece j in each i, $\sigma_{i,j}^2$, is computed. The variances for each piece j in each period i may be computed and represented by $\sigma_{i\_Within}^2 = \sum_{j=1}^{i} \sigma_i^2$. For each i between 1 and L/2, the number of sample points in each piece i may be p. In other words, $p = L/i$. For each i between 1 and L/2, p sets of sample data, numbered with $q = 1$ to p, may be constructed, each of which includes samples for the same i value. The variance for each value of p may be computed as $\sigma_{i,p}^2$.

The variance of each piece p in each period i may be summed, represented as $\sigma_{i\_Between}^2 = \sum_{q=1}^{p} \sigma_q^2$. The value of i between 1 and $L/2$ that has the minimal value of $\sigma_{i\_Between}^2 / \sigma_{i\_Within}^2$ ay be selected as the period for the seasonality component. Alternatively, the Analysis of Variance (ANOVA) technique may be used to select the value of i. After the length of the period, i, is obtained, the value of the seasonality component of each period may be determined. The value of the seasonality component in each position q (where q varies between 1 and p) may correspond to the mean value of data points at the same position q in the p sample sets. We derive the value of seasonality component each period; the value of seasonality component in each position q is given by the mean value of data points at the same position q in the p sample sets:

$$S_{pi}(t) = \sum_{q=1}^{p} X_{qi}/p \qquad (3.9)$$

The burst component indicates a significant change from normal trend which is caused by external known or unknown factors. A burst in the model is defined as a suspicious resource Indicator or traffic indicators measured value exceeding a predefined threshold. First, we find the geographical scene for the given cell; determine the approximate interval for various possible events such as holidays, sports game, or assembly etc. The next step is to observe such burst regularly appears in a given fixed interval. If yes, we finally confirm if the given burst is a routine burst rather than an ad-hoc burst which is to be recognized as random spike in the next section. The routing bursts are recorded to a Burst Table for each cell. The table is a cell level burst distribution list to demonstrate when and how much burst should be imposed.

The random component can further be decomposed to stationary time series RS (t) and white noise RN(t). The measured value of LTE resource indicator or network traffic indicator minus the sum of the first three components is just the estimated value of random error component. The random error component at each busy hour is given by the mean value of random error component at each busy hour.

3.4   Application for RRC Connection Setups Evaluation

3.4.1   Data Preparation and Feature Selection

To test the model, we use RRC connection setup success rate, a typical KPI, to evaluate the performance of accessibility in the LTE network. According to 3GPP TS 36.331, RRC connection establishment is used to make the transition from RPC idle mode to RRC connected mode. The KPI-RRC connection setup success rate evaluates the RRC setup success rate with service related causes in a cell or cluster involved. It is highly correlated to a few typical LTE capacity indicators as follows. These eight LTE capacity indicators are utilized as the features in our study.

- Average number of connected users per cell
- Down link physical resource blocks (PRB) usage
- Up link physical resource blocks (PRB) usage
- Physical down link control channel (PDCCH) usage
- Paging resource usage
- Average number of active users in the up link buffer per cell
- Average number of active users in the down link buffer per cell
- Physical random access channel (PRACH) usage

The relation between RRC Connection Setup Success Rate and the LTE capacity indicators can be trained according to Eqs. 3.1, 3.2, 3.3, 3.4, 3.5 and 3.6. A sample data from 457 cells is randomly split into training set (70%) and test set (30%). The sample set includes 12,097 points with hourly granularity.

3.4.2   Derivation of Relationship Between LTE KPIs and Network Resources

Exponential model, sigmoid model and GAM model are applied to train the statistical relation between RRC Connection Setup Success Rate and the resource indicators. Mean Absolute Percentage Error (MAPE) is estimated to compare the accuracy and reliability of the three algorithms developed Eqs. 3.1, 3.2, 3.3, 3.4, 3.5 and 3.6.

Figure 3.6 shows Mean Absolute Percentage Accuracy (MAPA) for these three models, which is calculated as MAPA = 1 − MAPE. The MAPA of the Exponential Model, Sigmoid Model and GAM Model in the same training set is 85.58%, 88.25%

Fig. 3.6 MAPA of three models

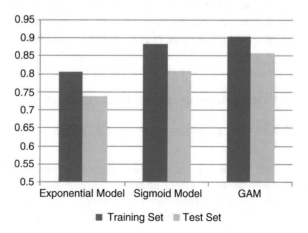

and 90.27% respectively. The sigmoid and GAM model have higher accuracy in training set than the exponential model. The primary reason of the lower accuracy for exponential is that more than 56% of the sample data in training set stay in Zone 1 and Zone 2. So, the exponential model does not fit well for Zone 1 and 2 where the gradient of LTE KPI is relatively smooth. On the other hand, the GAM and sigmoid model presents a fitting wellness in Zone 1 and 2. Between GAM and sigmoid model, the accuracy in training set does not make big difference. However, the difference is seen in test set in which the accuracy of sigmoid model is 80.77% while that of GAM is 85.67%. The higher accuracy of GAM can be explained as: sigmoid model is a centrosymmetric function with an inflection point, the cliff jumping point, at middle point of the max minus the minimal value of resource indicator. However, the fact is cliff jumping spot is normally left skewed rather than staying at middle point in the resource indicator dimension. This is well explained that some typical LTE KPIs such as Call Drop Rate and RRC connection setup success rate will not degrade until 50% of the resources are consumed.

Figure 3.7 also explains this situation. The RRC Connection Setup Success Rate starts to degrade when a few resource indicators reaches 20–40% range rather than 50%. Since it is a locally fitting algorithm, the GAM model can overcome this weakness rather than fixing the cliff jumping spot to be the point of 50% of resource indicators. Since there are 8 resource indicators in our model, which means it needs 8 dimensions to fit RRC Connection Setup Success Rate by GAM. Figure 3.7 visualizes the 3D relation between any two resource indicators and RRC Connection Setup Success Rate while setting the value of the other 6 dimensions to be their individual mean value.

In testing the significance of each smooth term, the Resource Indicator Average number of active users in the uplink buffer per cell shows a non-significant result, 0.115624. All of other resource indicators are significant at 95% confidence level. The resource indicator in Fig. 3.7 also shows a pattern against domain knowledge which is RRC Connection Setup Success Rate should drop when there are more active users in the uplink buffer. However, Fig. 3.7 shows that average number of active users in the uplink buffer is positively correlated to RRC Connection Setup Success Rate. Hence, a further step is conducted to utilize Analysis of Variance (ANOVA) to compare the existing model (Model A) and model B in which the average number of active users in the uplink buffer is deleted in Table 3.1.

The testing result indicates that the deletion of this resource indicator caused a highly significant increase in deviance emphasizing the fact that deletion is a better test than inspection of parameters. Eventually, we test the MAPA for Model A (all 8 resource indicators) and Model B (excluding average number of active users in the uplink buffer) with GAM. The accuracy rate in both is improved from 90.27% to 92.83% in training set and from 85.67% to 88.95% in test set. The result shows that the GAM, which excludes average number of active users in the uplink buffer, is the optimal model to evaluate RRC connections setup success rate in LTE.

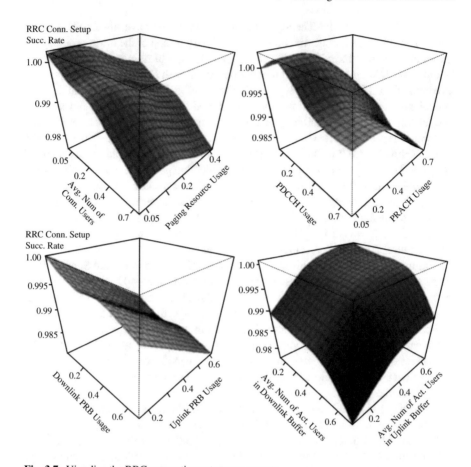

Fig. 3.7 Visualize the RRC connection setup success rate

Table 3.1 ANOVA test result

Model	Resid. Df	Resid. Dev	F Value	Pr(F)
Model A	95.99	29694.72	3.9544	0.0051
Model B	101.01	34493.71		

3.4.3 Forecasting RRC Connection Setup Success Rate

After deriving the relation between RRC Setup Success Rate and the relevant resource indicators, we verify the forecasting model. The test uses the same training data to construct the forecasting model for each resource indicator. The MAPA in training set and testing set, and the R2 (goodness of fit) are the 3 parameters to verify the accuracy and reliability of the forecasting model.

Table 3.2 shows the testing result, where the MAPA in training set ranges from 88.42% to 95.63% with a mean value 92.57%. It indicates the fitted values of the resource indicators in their respective training set shows very low variance between

Table 3.2 Testing result of forecasting model

Resource indicator	MAPA training set	MAPA testing set	R^2 (Goodness of fit)
Average number of connected users	0.9563	0.9022	0.9267
Downlink PRB usage	0.9481	0.8944	0.9103
Uplink PRB usage	0.9125	0.8273	0.8871
PDCCHU usage	0.9204	0.8617	0.9011
Paging resource usage	0.9101	0.8459	0.8956
Average number of active users in the downlink buffer	0.9486	0.9075	0.9302
PRACH usage	0.8842	0.8393	0.8699
Mean value	0.9257	0.8683	0.9029

true and fitted values. The MAPA in testing set also performs well. The MAPA of testing set ranges from 82.73% to 90.75% with a mean value of 86.93%, which is not far from the mean value in training set 92.57%. This indicates the reliability of the model is acceptable since the MAPA does not drop much from training to test set. The goodness of fit of all the resource indicators are above 0.8699, which is also a optimal result showing the independent variables in terms of those four components can well explain the variance of each resource indicator.

The MAPA of all resource indicators in both training and testing set are acceptable. The R square per resource indicator also indicates that the variance of each resource indicator can be well explained. At last, we pick up a resource indicator to visualize its test results in training set and testing set. The result also shows that the true values and fitted values of all the 457 cells in a given busy hour (8 pm) in a given day (Monday) are most overlapped both in the training set and testing test.

References

1. Tsao, Shiao-Li, and Chia-Ching Lin. (2002) "Design and evaluation of UMTS-WLAN interworking strategies." Vehicular Technology Conference.
2. Szlovencsak, Attila, et al. (2002) "Planning reliable UMTS terrestrial access networks." Communications Magazine 40.1 (2002): 66–72.
3. Ricciato. F. (2006). Traffic monitoring and analysis for the optimization of a 3G network. Wireless Communications. IEEE. 13(6). 42–49
4. Ouyang. Ye. (2012) Traffic Dimensioning and Performance Modeling of 4G LTE Networks. Diss. Stevens Institute of Technology.
5. Amzallag, David, et al. (2013) "Cell selection in 4G cellular networks." Mobile Computing 12.7 (2013): 1443–1455.
6. Navaie. K., Sharafat, A, (2003) "A framework for UMTS air interface analysis". Electrical and Computer Engineering, Canadian Journal of. Vol. 208. no. 3/4. pp.113.129.
7. Engles. A., Rever. M., Xiang Xu, Mathar. R., Jietao Zhang, Hongcheng Zhuang, (2013) "Autonomous Self-Optimization of Coverage and Capacity in LTE Cellular Networks", Vehicular Technology, Vol.62. no.5. pp.1989.2001.

Chapter 4
Trending Device Readiness and Device Return Rate

The usage of mobile devices has significantly grown in recent years, and they have become an integral part of our daily lives. They are being designed and manufactured at a rapid rate to satisfy market demand. Companies manufacturing mobile devices and wireless network providers marketing mobile devices may periodically review device return rates and causes of returns after the devices are launched to market. A clear and correct review of device return rate would help the manufacturers or the wireless network providers to determine success of a device model or make alterations to the subsequent design. Like the device return rates, the wireless network operator may also monitor a device's manufacturing progress to determine whether the device would be available to users based on a particular schedule. Organizations need to determine whether or not a device is ready to be launched into a consumer market.

So far however, neither the manufacturer nor the wireless network provider is able to predict or forecast the device return rates and their causes. In addition, no organization is able to quantitatively identify maturity of device readiness for market launch.

In this chapter, a model is introduced to forecast the device return rate and identify the primary root causes of a return. Such forecasting (or prediction) could enable supply chain and manufacturing enterprise to better track the trend of device quality, estimate inventory, estimate backup/replacement devices, determine how many pre-launched devices to be ordered from a manufacturer, balance Certified Like New Return (CLNR) and new devices, and optimize the resource. The model can be constructed to forecast (or trend) device return rate by using one or more algorithms. It is not limited to mobile devices and can be applied to other domains where forecasting return rates is needed. It also can be generalized to resolve issues related to supply chain stocking and returns.

Another model is also presented in this chapter to evaluate device readiness or maturity for a pre-launched device and forecast the time to market for the pre-launched device. It helps device manufacturers and commercial customers to monitor progress of device development for the pre-launched device. The model

Y. Ouyang et al., *Mining Over Air: Wireless Communication Networks Analytics*, https://doi.org/10.1007/978-3-319-92312-3_4

forecasts device maturity, which can be applied to other fields such as product maturity, software maturity, application quality maturity, etc. It provides a generalized model applicable to resolve quality maturity related problems. Furthermore, the model can classify Original Equipment Manufacturers (OEMs) based on respective readiness curves associated with devices manufactured by the OEMs. According to classified clusters, OEMs sharing an identical classification can have similar patterns of quality maturity in developing a device. Such classification can help wireless network providers estimate device readiness for a given OEM more accurately as well as rate OEMs based on manufacturing performance. Wireless network providers can manage OEMs during device development in a well-targeted manner. Furthermore, the wireless network providers may leverage the disclosed implementations to estimate device readiness for pre-launched devices OEMs. This model provides a neutralized and fair scheme to identify device readiness.

4.1 Device Return Rate and Readiness Forecasting Strategies

We observed that correlations between return cause codes and corresponding return rates for one or more devices can be identified. Since a cause code of device return can represent a reason for return of a device, and a corresponding device return rate is indicative of a number of times the device is returned with the device return cause code. Based on this identified correlations, correlation indexes for each of the device return cause codes can be computed. The computed correlation indexes associated with each of the device return cause codes can be ranked. A subset of the device returns cause codes are then selected based on the ranking. The subset can include device return cause codes having higher computed correlation indexes relative to other device return cause codes. A relationship between the selected subset of the return cause codes and corresponding device return rates can be determined by using the subset of the device return cause codes. One or more measured key performance indicators (KPIs) of a particular device are then mapped to return cause codes of the device. A device return rate for the particular device can be estimated based on the mapped KPIs and the determined relationship.

The model representing a device's readiness for market based on one or more key performance indicators (KPIs) of the device can be initialized as a sigmoid curve, which is a mathematical function having an "S" shape. One or more KPI values may be fitted to the curve. Fitting is a process of associating a series of data points to a curve or mathematical function. Then, differences between measured value of KPIs of the device and values of KPIs fitted to the curve could be computed. An inflection point of the curve can be identified based on the computed differences. The inflection point represents a point on the curve at which a derivative representing a curvature of the curve changes sign. The shape of the curve is interpreted based on the identified inflection point, a readiness index and the curvature of the curve, where the device readiness index is based at least on a KPI that takes the most amount of time relative to other KPIs to cross a manufacturing performance threshold represented in the curve. Then, a state of readiness of the device can be determined based on the interpreted shape of the curve. The state of readiness of the device can represent a state of readiness for launch to a consumer market.

The model for device return rate forecasting can provide a report describing the forecasted return rate, which may be used to modify aspects related to a supply chain or an inventory of devices. Since the rate of return for a device for a cause code can be forecasted in advance of actual returns and before the device is launched to the market, the manufacturer or a wireless network provider would be able to take pre-emptive actions to address issues with the device.

The model for device's readiness can also provide a report describing the determined state of readiness of the device, including a graphical representation of the curve, a readiness index and KPIs considered in determining a state of readiness of the device. The report includes a performance classification of a manufacturer of the device that can be determined based on the interpreted shape of the curve. It could be used to modify aspects related to a manufacturing and/or supply chain. Since the readiness of a device can be determined before the device is launched to market, the manufacturer can take pre-emptive actions to address issues with the device so that the device may be released to the market in a timely manner.

4.2   Device Return Rate Forecasting and Readiness Forecasting Model

4.2.1   Mobile Communication Service for Forecasting Model

The Fig. 4.1 illustrates a system which offers a variety of mobile communication services associated with forecasting device return rates and causes, and evaluating device readiness. MS in the Fig. 4.1 represents mobile station for which the device return rates and causes, and readiness could be predicted. Such mobile stations could be any form of portable handsets, smart-phones or personal digital assistants. Program applications which assist in forecasting device return rates and causes, and readiness can be configured to execute on these MSs. The mobile traffic network provides mobile wireless communications services to these mobile stations, as well as other mobile stations, via a number of base stations (BS). The mobile traffic network allows users of the mobile stations to initiate and receive telephone calls with each other, as well as telephone stations through the public switched telephone network or PSTN. The network also offers a variety of data services via the Internet, such as downloads, web browsing, email, etc.

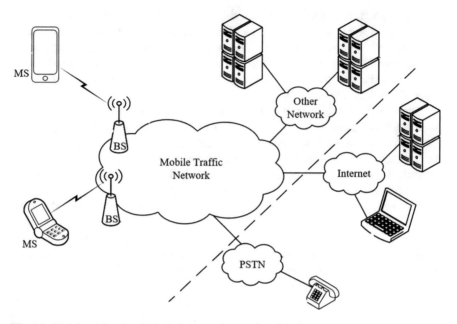

Fig. 4.1 High-level functional block diagram of network and device

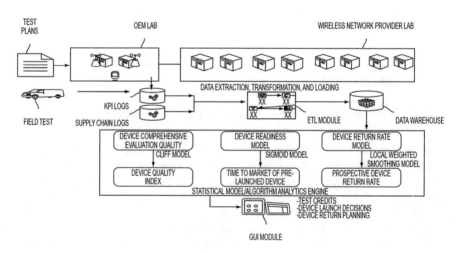

Fig. 4.2 Framework to obtain and store operational parameters

4.2.2 Parameters Obtaining and Storage

The framework which obtains and stores operational parameters related to mobile stations is shown in Fig. 4.2. This figure includes the Test Plans, Original Equipment Manufacturer (OEM) Lab, Wireless Network Provider Lab, KPI Logs, Extract-

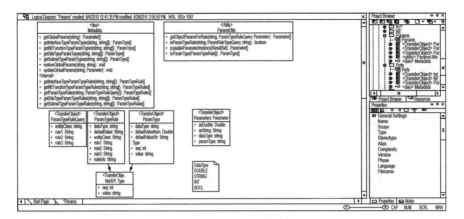

Fig. 4.3   Interface to define relations between data attributes in the metadata

Transform-Load (ETL) module, Analytics Engine, Graphical User Interface module (GUI) module, Field Tester, Supply Chain Logs, and Data Warehouse. Organization, such as wireless network operator, could operate the framework to determine the quality of mobile devices that have been manufactured by certain manufacturers for customer of the wireless network provider.

Specifically, the test plans may include data and associated guidelines on how to test the mobile station for quality. Testing mobile station could be performed at OEM Lab which may be a testing facility operated by the manufacturer of the mobile station. The general purpose of OEM lab and wireless network provider lab can be to perform one or more tests or measurements on the mobile station to determine operational parameters associated with the mobile station. KPI logs include data from field tester and OEM lab, which include KPI data, mean time between device failure, mean time to device repair, and any other performance parameter. The data from the KPI logs and the supply chain logs can be retrieved or extracted by the ETL module. The ETL module may extract, transform and load transformed data into data warehouse. The metadata in the data warehouse can define data attributes as well as their relations, and may include two types of metadata: Performance data attribute and a configuration data attribute. Performance data attributes include device KPI name, device KPI unit, device KPI threshold (max and limit value), wireless network (RF) KPI name, RF KPI unit, RF KPI threshold (max and limit value), etc. Configuration data attributes include device name, OEM name, device type, hardware configuration parameters, software parameters, sales data, returns data (per cause code), etc.

Once data attributes are defined in a metadata file, their relations can be defined. Figure 4.3 shows an interface to define the relations between data attributes in the metadata. The interface can be web-based interface and allows the configuration of mappings between both standard and proprietary data formats, as well as customizing the conversion of data types.

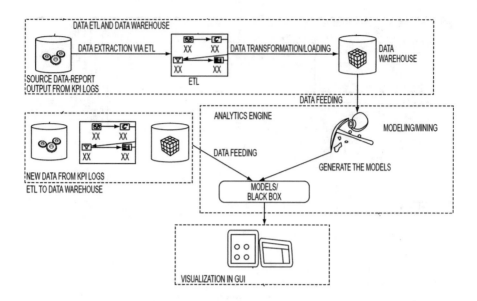

Fig. 4.4   Incremental learning in analytics engine

4.2.3   Analytics Engine

The analytics engine in the framework includes one or more processors, storage and memory to process one or more algorithms and statistical models to evaluate quality of the mobile station. It may train and mine the data from ETL module. The training set can be a set of data used to discover potentially predictive relationships. It can be used in artificial intelligence, machine learning, genetic programming, intelligent systems, and statistics. A training set can be implemented to build an analytical model, while a test (or validation) set may be used to validate the analytical model that has been built. Data points in the training set may be excluded from the test (validation) set. Usually a dataset is divided into a training set and a validation set (and/or a 'testing set') in several iterations when creating an analytical model. Open interfaces (e.g., APIs) may be provided to vendors for reading/writing data between the ETL module and the analytics engine and for visualizing analytics results between the analytics engine and GUI.

Data can be processed incrementally by the analytics engine for instantaneous learning as shown in Fig. 4.4. Incremental learning is a machine learning paradigm where a learning process takes place whenever new example(s) emerge and adjusts what has been learned according to the new example(s). Incremental learning differs from traditional machine learning in that incremental learning may not assume the availability of a sufficient training set before the learning process, but the training examples appear over time. Based on this paradigm, the algorithms utilized by the analytics engine may be automatically updated by re-training the data processed by

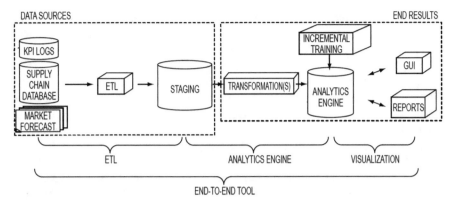

Fig. 4.5   Data sources of an extract-transform-load module and analytic engine

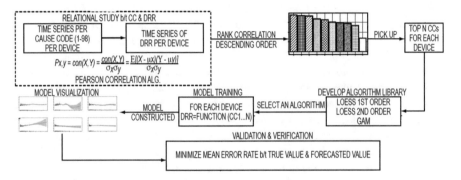

Fig. 4.6   Framework forecasting device return rate and cause of return

the analytics engine. A dynamic sliding window method may be used to provide data from the ETL module to the analytics engine for algorithm training by the analytics engine. The dynamic sliding window may be used by the ETL module to incrementally provide, for example, operational parameters from the mobile station to the analytics engine.

The analytics engine could have different data sources as illustrated in Fig. 4.5. There can be three primary data sources to ETL module and the analytics engine, including KPI logs, supply chain database, and market forecast data.

4.3   Implementation and Results

4.3.1   Device Return Rate Forecasting

The framework to forecast device return rate and cause of return is illustrated in Fig. 4.6, and is implemented in the analytics engine we discussed in the previous section in this chapter. Correlation module can identify correlations between device return cause codes and corresponding device return rates for one or more devices, and also can identify the correlation between a time series per cause code and a time series per device return rate. Based on the identified correlations, the correlation indexes for each of the device return cause codes can be computed by the module. The computed correlation indexes associated with each of the device return cause codes can be ranked by ranking module. A subset of device return cause codes is then selected based on the ranking by the selector module. The subset can include device return cause codes having higher computed correlation indexes relative to other device return cause codes.

After obtaining the correlation index per cause code, they may be ranked in a descending order. Then, the top "N" cause codes can be selected by the selector module as independent variable candidates in equations used by the relationship and algorithm computer to forecast device return rate. A device return rate for the particular device can then be estimated (or forecasted) by the model training and forecasting module based on the mapped KPIs, the relationship between the selected subset of the return cause codes and the corresponding device return rates previously determined by the algorithm computer. Visualization module may generate a model visualization indicative a mapping between the measured key performance indicators (KPIs) and return cause codes of the device. Validation and verification module can validate a difference between an estimated device return rate and a true (or measured) value of a device return rate.

Figure 4.7 illustrates a map-able relation between measured KPI and cause codes. To forecast the potential return rate for a given pre-launched device, correlation module can map the measured KPIs to cause codes due to the unavailability of cause codes data for a pre-launched device. KPIs belonging to certain category are mapped to each corresponding cause code. Then, training and forecasting module leverages the formula derived from relationship and algorithm computer above to derive the relation between measured KPIs and potential device return rate. Figure 4.8 displays the correlation index per primary cause code in a descending order. Such a relational graph can be an output from selector module, which prepares independent variable

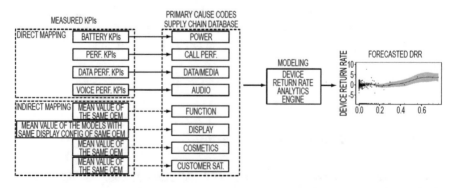

Fig. 4.7 Relations between measured key performance indexes and cause does

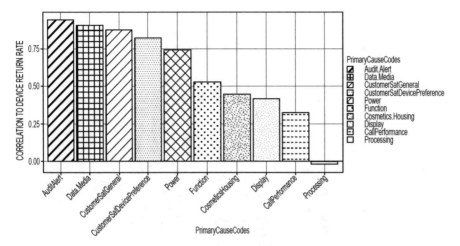

Fig. 4.8 Correlation index per primary cause code in descending order

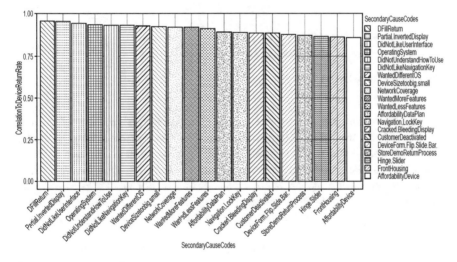

Fig. 4.9 Correlation index per secondary cause code in a descending order

candidates to regress (or forecast/estimate) device return rate. Graph illustrates primary cause codes on the horizontal axis and correlation to device return rates on the vertical axis. Figure 4.9 displays the correlation index per secondary cause code in a descending order. Such a correlation index can be output from selector module, which prepares the independent variable candidates to regress device return rate. Figure 4.10 illustrates which cause codes are the primary causes of return for a given device type. Such a word cloud may be generated and displayed by the visualization module in addition to the forecasted return rate report discussed above. A word cloud is a visual representation for text data. The frequency of

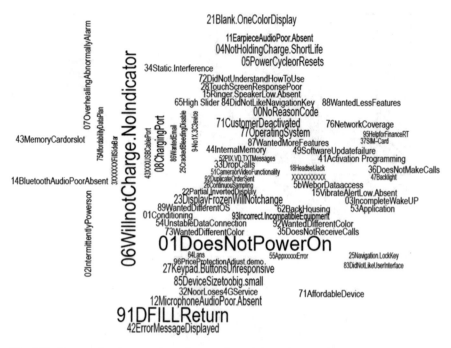

Fig. 4.10   Cause codes that are primary causes of return for a given device type

occurrence return cause codes may be used to determine the size of the text associated with the cause code. In other words, when visualized text size of a particular cause code is greater with respect to other cause codes, this indicates that the particular cause code more frequently is a reason for a device's return relative to other cause codes. Word clouds are useful for quickly perceiving the most prominent terms. Such visualization can help a wireless network provider team quickly identify the primary causes of return for a given device (e.g., the mobile station).

Figure 4.11 visualizes relations between any selected cause code and device return rate. It shows that the selected cause codes jointly regression device return rate in a multi-dimensional space. The curves illustrated in this figure can be computed by the analytics engine using a set of data points corresponding to device return rates. The visualizations of Fig. 4.11 can be considered to be scatter-grams. In a scatter-gram, when a parameter exists that is systematically incremented and/or decremented by the other, the parameter is called the control parameter or independent variable and is customarily plotted along the horizontal axis. The measured or dependent variable is customarily plotted along the vertical axis. If no dependent variable exists, either type of variable can be plotted on either axis or a scatter-gram can illustrate a degree of correlation between two variables. As an illustrative example, each smoothed curve value can be computed by the analytics engine using a weighted quadratic least squares regression over the span of values of the

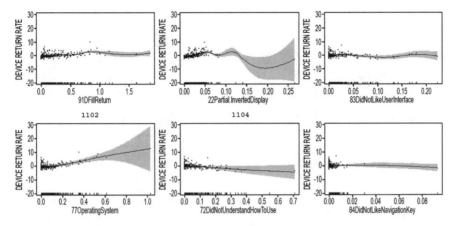

Fig. 4.11 Relations between selected cause codes and device return rates

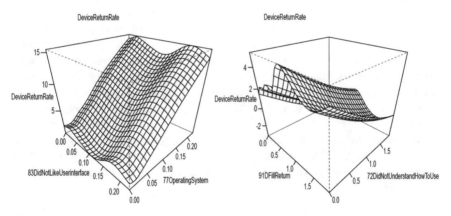

Fig. 4.12 Relations between two selected causes codes and device return rates

y-axis scatter-gram independent variable. Figure 4.12 visualizes relations between any two selected cause codes and device return rates in a multi-dimensional space. The surfaces can be computed by the analytics engine from a combination of any two visualizations illustrated in Fig. 4.11. The surfaces can be computed using multiple scalar variables and associated the variables for different axes in phase space. The different variables may be combined to form coordinates in the phase space and displayed as surfaces.

4.3.2   Device Readiness Forecasting

The framework evaluating the device readiness is illustrated in Fig. 4.13. The model representing the readiness of a device for market based on one or more key

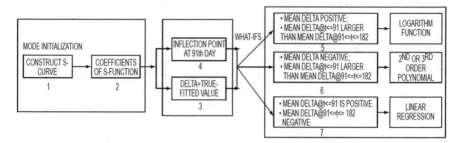

Fig. 4.13 Framework to evaluate device readiness

performance indicators (KPIs) can be represented by a curve, such as an S-curve or a sigmoid curve. Once the model is initialized, the model computer computes coefficients of a sigmoid function (e.g., using a Least Square Method) based on one or more key performance indicators (KPIs). "Least squares" indicates that an overall solution minimizes the sum of the squares of the errors made in the results of each equation of the sets of equations. To initialize the model, model computer may assume that device readiness presents a sigmoid or S-shape relation to time. As time progresses, more testing data becomes available. The model shape may be adjusted by model computer from sigmoid to other shapes. The threshold to change model shape may be triggered by several conditions, including, but not limited to, mean error rate, a determination of whether the curve is concaved up or down, and curve curvature etc.

The model for device readiness forecasting utilizes one or more algorithms to determine device readiness. It assumes that the bench mark value of a given KPI for a device is the measured value at Day 0 (or the first day) of the manufacturing cycle of the device.

$$KPI_{BencheMark} = KPI_{Measured@Day0} \qquad (4.1)$$

For this particular KPI, the model can specify the KPI's acceptance value, which is the threshold to pass this KPI. It may be specified by the wireless network provider to the OEM manufacturing the device. Thus, it can be specified that $KPI\_Acceptance = p$. The next step is to define a readiness index for a KPI of a device. The model specifies that the readiness index for a KPI at a given day t is given by the absolute difference of KPI measured at day t and an absolute difference of a KPI acceptance value and a KPI benchmark. The KPI acceptance value is a KPI value that is acceptance to a wireless network provider to whom the device is to be delivered by the OEM. The KPI benchmark value is a recommended or benchmark KPI value for the particular device. The model may compute the Device Readiness Indicator (DRI) as:

$$ReadinessIndex_{KPI} = \begin{cases} \left| \dfrac{KPI_{Measured@Day\_t} - KPI_{BM}}{KPI_{Acceptance} - KPI_{BM}} \right| & when \quad KPI_{Day\_t} \geq KPI_{BM} \\ 0 & when \quad KPI_{Day\_t} < KPI_{BM} \end{cases}$$

$$(4.2)$$

$KPI_{Day\_t} \geq KPI_{BM}$, indicates that the KPI measured at day t is better than the KPI benchmark value.

The standard sigmoid function is represented as Eq. 4.3. However, the sigmoid curve of Time To Market Maturity (TTMM) model may not correspond precisely to the standard sigmoid curve. Hence, a DRI function can be represented as:

$$y = f(x) = \frac{1}{1 + e^{-x}} \qquad (4.3)$$

$$ReadinessIndex_{KPI} = \frac{A_1}{1 + e^{-B.(t-C)}} = \frac{A}{1 + e^{B(C-t)}} \qquad (4.4)$$

where A denotes a maximum value of DRI, B denotes the curvature of the sigmoid curve. In a circular function, B can be termed as a phase of the sigmoid curve, and C denotes an inflection point of the sigmoid curve. In this model, if the maximum value of the DRI is known to be 1, we have $ReadinessIndex_{Max} = 1$, and get $A = 1$. Assuming that the time between when a device is safe for network and when testing is complete is assumed to be 182 days. The inflection point can be determined to be the 91st day. Accordingly, we have Eq. 4.5, and get Eq. 4.6 after transforming:

$$RI_{KPI} = \frac{1}{1 + e^{-B.(t-91)}} \qquad (4.5)$$

$$t = 91 - \frac{\ln\left(\frac{1}{RI_{KPI}} - 1\right)}{B} \qquad (4.6)$$

Both t and B are nonnegative in this model, we assume that $RI_{KPI} = 0$ and $t = 0$. It can be interpreted that at day 0 that DRI is also 0. When the model is initialized by the model, the model assumes that the testing work performed by the OEM in day 1 completes 1/182 of overall testing work, which is represented by 1/182 of DRI. Therefore, assuming that $RI_{KPI} = 1/182$, we have:

$$t = 91 - \frac{\ln\left(\frac{1}{1/182} - 1\right)}{B} = 91 - \frac{5.198}{B} \qquad (4.7)$$

Thus, $t = 91 - (5.198/B) = 1$ day. Accordingly, it can be determined in this example that $B = 0.5776$. Finally, the TTMM model can be represented by the function group below and Eq. 4.2:

$$t_{KPI} = 91 - \frac{\ln\left(\frac{1}{RI_{KPI}} - 1\right)}{0.5776} \tag{4.8}$$

$$TimeToMarket_{KPI} = Day\_x_{KPI} + (182 - t_{KPI}) \tag{4.9}$$

The model could perform the following steps to construct the model. At first, the model may compute a DRI value for a particular KPI_i. The DRI value for the particular KPI_i may be computed as:

```
for kpi from 1 to i {
Get i-th kpi's benchmark value KPIi_BM;
Get i-th kpi's acceptance value KPIi_Acceptance;
Get i-th kpi's measured value at DayX KPIiMeasured@Dayx;
if KPIi Measured@Dayx >= KPIi_BM{
ReadinessIndex_kpii = abs((KPIi_Measured@Dayx - KPIi_BM)/KPIi_Acceptance-
KPIi_BM)
} else {
ReadinessIndex_kpii = 0;
}
}
```

Once the model has computed a DRI value for a particular KPI_i, the model can compute $tKPI$ and days to market for KPI_i. The $tKPI$ and days to market for KPI_i can be computed as:

for kpi from 1 *to i*
{

$$ReadinessIndex_{KPI} = \begin{cases} \left| \dfrac{KPI_{Measured@Day\_t} - KPI_{BM}}{KPI_{Accep\tan ce} - KPI_{BM}} \right| & when \quad KPI_{Day\_t} \geq KPI_{BM} \\ 0 & when \quad KPI_{Day\_t} < KPI_{BM} \end{cases}$$

$$t_{KPI} = 91 - \frac{\ln\left(\frac{1}{RI_{KPI}} - 1\right)}{0.05776}$$

$$TimetoMarket_{KPI} = Day\_x_{KPI} + (182 - t_{KPI})$$

}

After tKPI and days to market are calculated for KPI_i, the model can select a maximum amount of time to market for KPI1 to KPI_i. The DRI value may be computed as:

for kpi from 1 *to i*
{
  $TimetoMarket = Max\{TimetoMarket_{KPI\_i}\} = Max\{Day\_x_{KPI\_i} + (182 - t_{KPI\_i})\}$
}

After creating the forecasting model, Fig. 4.14 illustrates the readiness curves LARK, NIGHT OWL and REGULAR BIRD associated with different OEMs. The vertical axis indicates Device Readiness Indicator (DRI) values. The horizontal axis

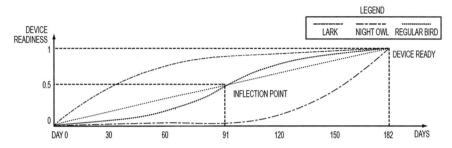

Fig. 4.14 Readiness curves

is a temporal axis and indicates days in a manufacturing cycle of a device (e.g., the mobile station). Curve fitting is a process of constructing a curve or mathematical function that has the best fit to a series of data points, possibly subject to constraints. Curve fitting can involve either interpolation, where an exact fit to the data is required, or smoothing, in which a "smooth" function is constructed that approximately fits the data. Inflection point identifier can compute differences between each true value and fitted value (obtained by a sigmoid curve). An inflection point of the curve can be identified by the inflection point identifier based on the computed differences by the inflection point identifier. The inflection point represents a point on the curve at which curvature of the curve changes sign. As an illustrative example, the inflection point of an S-curve may be identified as a point corresponding to the 91st day out of 182 day manufacturing cycle of the mobile station.

Chapter 5
Evaluating VoLTE Voice Quality in LTE Networks

Voice communication is largely used nowadays by LTE phone users, despite introduction of many Apps on the market. Therefore, mobile voice call quality assessment is still an important metric to evaluate for mobile carriers. Research in assessing speech and voice quality mainly focuses on audio clips analysis. POLQA (standing for Perceptual Objective Listening Quality Assessment) is the standard for voice quality evaluation, taking audio clips as the input and giving an objective quality evaluation.

The main drawbacks of such evaluation methods through audio feature analysis it not scalable and diagnosable. In fact, testing results are based on human perception and are subjective to language. Furthermore, the evaluation is considered separately without considering the wireless network itself as a causal factor. In this chapter, we provide a guideline for network diagnose and optimization through another perspective. The main idea is to collect data as the source of knowledge to leverage the causal factor that directly impacts the mobile voice quality. Collection of data is done through crowdsourcing, using a great number of data provided by user and experiencing a broad range of voice quality (depending of the strength and interferences of the signal).

This crowdsourcing allows connecting voice quality with the different network indicators available from traffic data. The calibration is performed once with POLQA standard, which is the only step where POLQA is needed. From the machine learning side, clustering and regressions are needed to adjust the connection between radio frequency (RF) indicators, network indicators and voice quality. In the final section, a pilot trial based on a crowdsourcing App is presented, showing that this methodology does not require additional hardware or human effort and provides an high model accuracy with a strong diagnosability.

© Springer International Publishing AG, part of Springer Nature 2018   69
Y. Ouyang et al., *Mining Over Air: Wireless Communication Networks Analytics*,
https://doi.org/10.1007/978-3-319-92312-3_5

5.1   POLQA in Assessment of Voice Quality

5.1.1   POLQA Standard

Most published research findings in the quality assessment of mobile voice call mainly focus on evaluating the audio quality of the speech. In these methods, Perceptual Objective Listening Quality Assessment (POLQA), an ITU-T standard (see Fig. 5.1), is widely utilized to provide objective voice quality evaluation through signal analysis. The POLQA specially supports new types of speech codecs used in 3G and 4G LTE networks. It takes audio clips as inputs and compares the clips with pre-recorded referential speeches to rate a degraded or processed speech signal in relation to the original signal. The difference between the two signals is counted as distortions. The accumulated distortions in the speech files are scored from 1 to 5 based on Mean Opinion Scores, and such score is the qualified assessment of the input audio quality. In this method, to evaluate the quality of mobile voice for cellphones, each test phone needs to connect to a POLQA box, which includes the POLQA assessment algorithm, microphone, audio recorder, playback, etc. Each POLQA box originates phone calls to other phones, plays the prerecorded reference audio clips, and record the received audio signal. The recorded audio clips are then processed inside the box through the POLQA algorithm to calculate the quality score.

Under the architecture of POLQA, researches in assessing speech and voice quality mainly focus on audio clips analysis [1–6], human voice modeling [7], and language processing [9, 10]. Subjective voice quality assessment is conducted to consider human perception as a key factor to evaluate the clarity and understandability of the voice call [8]. To further assess the quality of human voice, features in different languages and tones are extracted and modeled [11].

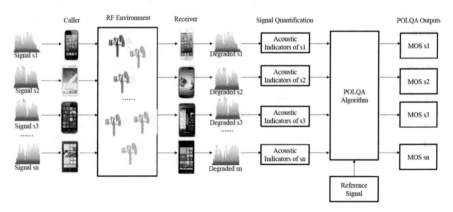

Fig. 5.1 Voice assessment using POLQA

5.1.2 *Scalability and Diagnosability in Assessment of Voice Quality*

However, such methods suffer from high expense, special devices, and extremely limited purpose and usage. First, evaluating audio features using these methods is cumbersome and costly, which requires noiseless testing environment and professional hardware, such as high definition recorders, headphones, and playbacks. Moreover, the testing results usually are only subjective to languages and tones. The evaluation of audio quality is not able to analyze the root cause of the change of the quality, which means it cannot provide a guideline for mobile system diagnose and optimization. More importantly, they cannot be directly applied in large-scale mobile network for network-side voice quality assessment and issue localization.

To overcome these limitations, a feasible voice quality assessment method needs to involve as less hardware and human efforts as possible, and must be interpretable and can be directly mapped to network indicators. These requirements make the method must be scalable and diagnosable.

5.2 CrowdMi Methodology

We observed that the major cause of the mobile voice degradation is nothing else but the signal propagation in wireless environment. Based on this observation, a wireless analytics algorithm is designed to model the mobile voice quality by mining various types of network indicators. Such model does not directly measure the audio features. It builds a quantified causal relationship between the change of voice quality and the deviation of the network conditions, through large-scale data crowdsourcing from users in different network scenarios.

The wireless analytics model, named CrowdMi in this chapter, models the mobile voice quality by mining various types of network indicators and respects POLQA as the standard for quantifying mobile voice quality. It collects both voice audio clips and network indicators during the phone call tests, classifies these testing data into different groups based on different radio frequency (RF) features, performs clustering on each group based on the relevant network indicators which impose heavy impact on the network performance, and eventually creates a predication model via regression. CrowdMi consists of two phases: Training and testing phases, and can be installed as an App in smartphones to perform voice quality assessment. To collect data, phones installed with the CrowdMi make voice phone call to each other, and also collect data including audio clips and network indicators during the phone call.

5.2.1   RF Features Based Classification

The collected records are classified into different groups based on their RF quality. Since mobile voice quality are mainly impacted by two facts: Coverage and interference. We consider Reference Signal Received Power (RSRP) and Signal-to-Interference-plus-noise ratio (SINR) as the two main features for the RF quality to classify the records.

5.2.2   Network Indicators Selection and Clustering

We then perform clustering to cluster the records based on their network indicator inside each RF group. There are hundreds of kinds of network indicators in mobile network, but only a few of them may impact the voice quality. We design a feature selection method to select the most relevant network indicators and then use them as features to perform clustering, which clusters the data to several groups based on the availability, sufficiency, and assignability of the network resource. The selected network indicators are treated as features for the clustering. We apply K-Medoids method to do the clustering work by imposing a new converging rule to identify the best K to converge the model. This is because the network indicator consumption shows high deviation due to peak time and rush hours. The network indicators, containing many spikes and outliers, may be diluted in K-Means and other similar algorithms.

5.2.3   Association Between Network Indicators and POLQA Scores

After clustering all the records into different clusters, for each cluster, we associate the selected network indicator to the computed POLQA scores by regressing network indicators that have high correlation to voice quality. We propose Adaptive Local Weight Scatterplot Smoothing (A-LOESS) regression, which improves original LOESS [13] by adding adaptive window size to regress the features and computing the estimated voice quality score.

5.2.4   Model Testing

Once the voice quality model has been built, we no longer rely on the POLQA and have a full leverage of the mode. To assess the voice quality for a phone, CrowdMi only collects network indicator and RF data of each phone. Such data is treated as input for the model which assigns the RF quality group and network indicator cluster by measuring the similarity between the input records and the training model. The voice quality for this phone call could be calculated by the model using these features.

5.3   Technical Details on CrowdMi

5.3.1   Classification of Records

The RSRP refers to average power of resource elements that carry cell specific reference signals over the entire bandwidth. It is a direct cell signal strength indicator. Hence, RSRP is a representative indicator to denote coverage strength. On the other hand, a strong coverage cannot ensure a good RF quality. A strong covered area with high interference and noise may still degrade voice signal. SINR reflects the interference and noise condition, and is a typical indicator to represent interference condition. Normally, high SINR case always correlates to high RSRP unless in a high polluted RF environment. Furthermore, subject matter experts (M. [12]) have proposed scales of LTE signal strength for those signal indicators. Hence, we simply use the scales to classify RF quality and divide the records into different groups. Table 5.1 is the classification table based on the scales of RSRP and SINR respectively in CrowdMi.

5.3.2   Selection of Network Indicators

To do the feature selection, we design spatial silhouette distance (SSD) to measure the capability of each network indicator to differentiate the voice qualities, and only select the features with large SSD value. In the training dataset, we first divide all the records into different quality groups according to its POLQA voice score. Then, we calculate the SSD for each network indicator in each group, and use such value to determine the discrimination capability of the network indicator. More specifically,

Table 5.1 Classification on RF quality

Class No.	RSRP (dbm)	SINR (db)	Description
Class 1	>= -85	> 15	Good Cov. and Low Intf.
Class 2	>= −85	<= 15	Good Cov. and High Intf.
Class 3	(−105, −85)	> 15	Median Cov. and Low Intf.
Class 4	(−105, −85)	<= 15	Median Cov. and High Intf.
Class 5	<= −105	> 15	Poor Cov. and Low Intf.
Class 6	<= −105	<= 15	Poor Cov. and High Intf.

we follow ITU-T standard to divide the mobile voice data into four groups based on their POLQA score as follows: C1:[0,2), C2:[2,3), C3:[3,4), C4:[4.0,4.5].

Assume that each group C_k ($k = 1,2,3,4$) has n records and each records r^k_j has m network indicators. In each quality group C_k, for each network indicator point $R^k_{i,j}$ of each record r^k_j, we fist compute the Euclidean Distance (ED) to all the other points in the same group, and obtain the average intra-group ED $IntraED^k_{i,j}$, for this feature point. Then, for this feature point, we compute its ED to feature points in all the other groups and calculate the average inter-group ED $InterED^k_{i,j}$ for this network indicator point $R^k_{i,j}$. Following this way, we compute the average intra-group ED, and average inter-group ED for every feature point in the training records. After that, for each quality group, and each network indicator, we average out its intra-group ED over all the records inside the group to obtain its group-wised average intra-group ED $IntraED^k_i$, and obtain a group-wised average inter-group ED $InterED^k_i$ for this indicator. For each quality group C_i, the SSD for each indicator R^k_i is calculated by Eq. 5.1:

$$S^k_i = \frac{InterED^k_i - IntraED^k_i}{\max\{InterED^k_i, IntraED^k_i\}} \tag{5.1}$$

For each network indicator R^k_i in a quality group C_k, we use tri-cube weight function to weight it as follows:

$$W^k_i = \begin{cases} \left(1 - \left|\sum_{i,j=1}^{4}\left(R^k_i - R^k_j\right)\right|^3\right)^3 & if \left|\sum_{i,j=1}^{4}\left(R^k_i - R^k_j\right)\right| < 1 \\ 0 & if \left|\sum_{i,j=1}^{4}\left(R^k_i - R^k_j\right)\right| \geq 1 \end{cases} \tag{5.2}$$

Finally, we obtain the SSD for each network indicator R_i over all the RF groups:

$$S_i = W^k_i \times S^k_i \tag{5.3}$$

After we obtain the SSD for all the network indicators, we select indicators whose $S_i = 0.7$ as the features to perform clustering, considering they are the discriminative features and are highly correlated to the voice quality.

5.3.3   Clustering

We define an upper bound of cluster number u based on domain experience to choose the optimal number of clusters, k. We iterate k from 2 to u and perform K-Medoids clustering in each iteration. The optimal k is selected, such that the intra-cluster error is minimized and the inter-cluster distance is maximized as follows:

$$
\begin{cases}
0.7 \leq \dfrac{IntraSumOfError_{k+1}}{IntraSumOfError_k} \leq 1 \\[2ex]
0.7 \leq \dfrac{IntraSumOfError_{k+2}}{IntraSumOfError_{k+1}} \leq 1 \\[2ex]
0.7 \leq \dfrac{IntraSumOfError_{k+3}}{IntraSumOfError_{k+2}} \leq 1
\end{cases}
\tag{5.4}
$$

5.3.4   Regression

We pack the POLQA scores into different bins, and dynamically adjust window size for each local set by the distribution density of each bin. Based on domain experience in voice assessment in POLQA, we set 9 bins according to the POLQA scale: $Bin_0 = [0,0.5]$, $Bin_1 = (0.5,1)$, $Bin_2 = [1,1.5]$, ..., $Bin_8 = (4.5,5]$. We set an initial window width to 1/100 of range of sample points, and plot the scatterplot of all measured POLQA scores in an ascending order. Let f(x) denotes the scatterplot function, where x is from 1 to the number of POLQA sample points. First, for each Bina, we compute its distribution density by integrating the value of the scatterplot function in its range as follows:

$$
y_a = \int_{f^{-1}(0.5a)}^{f^{-1}(0.5a+0.5)} f(x)dx \quad (i = 0, \cdots, 8)
\tag{5.5}
$$

After that, we sort ya in ascending order. Let $S(y_a)$ min represent the bin with minimum y_a, $S(y_a)$ med represent the bin with median value of y_a, and $S(y_a)$ max represent the bin with maximum y_a, we dynamically calculate the window size by the sorting results, as follows:

$$
win\_size =
\begin{cases}
\dfrac{0.5 + 0.125 \cdot S}{100} \cdot N & if(S = 0, \cdots, 4) \\[2ex]
\dfrac{1 + 0.25 \cdot (S - 4)}{100} \cdot N & if(S = 5, \cdots, 8)
\end{cases}
\tag{5.6}
$$

Eventually, we use the adaptive window size calculated by the above equation to perform LOESS regression to the POLQA score based on the selected features.

5.4   CrowdMi Prototyping and Trial

5.4.1   Client and Server Architecture

The CrowdMi model consists of two major components: Client and Server as the Fig. 5.2 shows. The client is implemented as an App in Android smartphones, and its main functionality is to collect user data in different locations and network scenarios through crowdsourcing, and send back to the server. The server runs over wireless analytics algorithm, mines the collected data to build a model to assess the mobile voice quality based on the collected network indicator in the training phase, and takes the real-time data collected from the each of Clients and calculates the mobile voice quality for the client which is the current voice quality of the place where the Client locates in the testing phase.

The client in smartphones automatically monitors the network conditions of the phones and collects the data. In the training phase, each of such phones is operated by test engineers and connected with a POLQA box which includes several pre-recorded audio clips of reference speech, and a standardized objective voice quality measurement system that takes input voice clips, compares such speech with the reference speech, and calculate the quality of the voice. When training phase starts, phones with the client call each other, plays audio clips generated by the POLQA box, record the audio clips received from the other phone, and at the same time record the network conditions of the phone during the call. After each call ends, each POLQA box calculates the quality score of the recorded audio clips, and the client uploads the score and network indicators to the server. The server uses such data to build a model for voice quality assessment.

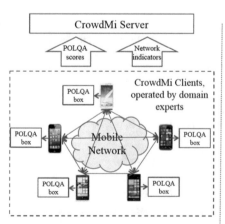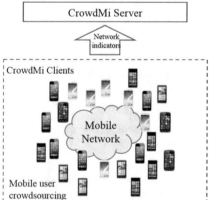

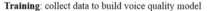

Training: collect data to build voice quality model | **Testing**: crowdsource data to assess voice quality

Fig. 5.2 The architecture of CrowdMi System

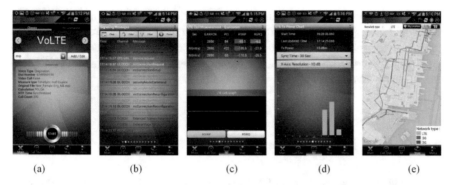

(a)   (b)   (c)   (d)   (e)

Fig. 5.3 User interface of the CrowdMi App running in LTE network scenario

Fig. 5.4 CrowdMi client in
training phase

In the testing phase, the phone with client needs not connect to POLQA box and
only runs the client. The client does not make phone calls, and runs in background to
collect network indicators of the phone. It sends data back to the server periodically,
reporting the network conditions of the phone in different locations.

Figure 5.3 shows the user interface of the client running in training phase in
VoLTE scenario. In Fig. 5.3, (a) is the front page of client, (b) shows log information, (c) is the visualization of network conditions, (d) displays visualization of
network conditions, (e) is the visualization of moving traces during data collection.
Such displays include rich log information and the visualization of assessment
results sent back from the server, such as KPIs, quality assessment scores, location
traces, etc. Such information can be displayed in real-time and is shown in various
types, which drastically facilitate the voice assessment of the network and help
domain engineers to diagnose the network issues. Figure 5.4 shows the client in
testing vehicle during training phase. The POLQA box needs not be connected, and
the visualization features can be turned off to run silently in the background when
doing the large-scale crowdsouring in the testing phase. The server builds a voice
quality model and assesses the voice quality of the cellular networks in different
locations and coverage conditions. In the training phase, the server collects data from

clients, and runs the mining algorithm to model the mobile voice quality using the received voice quality scores and network indicators. After the model is built, it is stored in the server. In the testing phase, for each of the client, the server periodically estimates the voice quality using the computed model, and such estimation is the voice quality assessment of the network in the client's location.

5.4.2   Pilot Trial and Results

We conduct a pilot trial in VoLTE network of different geographic areas with various network qualities to test accuracy of the voice quality model and evaluate the diagnosability of the system for finding relevant network indicators to the voice quality, lasting from December 2013 to August 2014. During the 9 months, we install clients app to 50 smartphones with Android 4.3 system which support VoLTE functionality in a major network carrier. The client measures all the needed network/ RF/device performance indicators, and collects and uploads test logs on a rotation basis. We select the pre-recorded List-11 Harvard Sentences of Female American English voices, each with 10 s length, as the audio input for the POLQA box. All the testing phones are in Time Synchronization Half-Duplex mode. When a phone calls another, and plays the audio clip, the receiver starts to compute POLQA score by comparing the received audio signal against the reference audio signal, and starts to play the same audio clip back to the caller at the same time.

Considering mobility is an important factor to voice quality, 77% are drive tests and 23% are stationary test. We randomly select wireless environment for each test case. We generate POLQA records in diverse wireless environment with different quality of coverage, interference, etc. In total, we collected 317 logs of POLQA test cases, where 299 are valid and 18 are error logs and thus discarded. The valid logs consists of 8987 POLQA voice records. All the records are classified into 6 groups by the measured RSRP and SINR values as Fig. 5.5 shows.

Table 5.2 shows the top 9 network indicators we selected by calculating SSD. The majority of the selected indicators are related to throughput and audio transmission which is under our expectation. To show the high correlation of the selected features to the voice quality, we utilize features of "RLC.DL.Throughput", "RTP.Audio.Rx. Throughput", and "Handover.Happening" in Table 5.2 and compute the POLQA scores on top of them. As expected, the POLQA score is high when the throughput indicators are high, and the score is low when handover happens frequently. The indicators are strongly correlated to the POLQA scores. To evaluate the accuracy of the A-LOESS algorithm, we use 75% of the data as training dataset, and rest 25% as testing dataset, and use Mean Absolute Percentage Error (MAPE) to calculate the error of the model as follows:

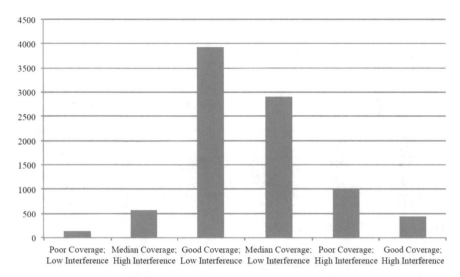

Fig. 5.5 Record distribution in RF groups

Table 5.2 Selected features

Features	SSD
MAC.DL.Throughput	0.8430
PDSCH.Throughput	0.8214
RLC.DL.Throughput	0.8186
RTP.Audio.Rx.Throughput	0.8057
PDCP.DL.Throughput	0.7934
RTP.Audio.Tx.Throughput	0.7928
RTP.Audio.Rx.Delay	0.7412
RTP.Audio.Rx.Jitter	0.7103
Handover.Happening	0.6974

$$e = \frac{1}{n} \sum_{i=1}^{n} \left| \frac{S_i^{POLQA} - S_i^{CrowdMi}}{S_i^{POLQA}} \right| \tag{5.7}$$

where S_i^{POLQA} and $S_i^{CrowdMi}$ are the voice quality score of the i-th record calculated by POLQA and the CrowdMi respectively. The MAPE in each RF group is shown as Fig. 5.6. From the Fig. 5.6, we can see that the MAPEs in the training dataset, except Group "Poor Coverage Low Interference", is less than 10%. The low MAPE in this group is not caused by the model, but insufficient records collected from the trial. It can be solved by conducting a few additional tests in the environments of such RF group. Overall, the MAPE maintains at a very low level, which indicates great model accuracy of the CrowdMi system.

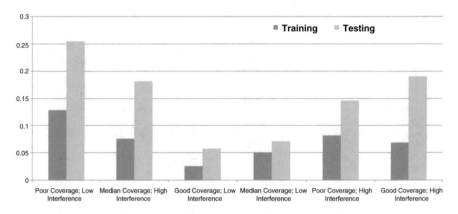

Fig. 5.6 MAPE of CrowdMi in each RF group

The reliability between training and testing set is robust as the difference of MAPE between training is small, which is no longer than 12.58%, also from Group "Poor Coverage Low Interference". It shows that the model is a valid approach which can be applied for POLQA assessment in the LTE network.

References

1. J. Berger, A. Hellenbart, R. Ullmann, B. Weiss, S. Mller, J. Gustafsson, and G. Heikkil, (2008) "Estimation of quality per call in modelled telephone conversations," Conf. Acoustics, Speech, and Signal Processing (ICASSP), Las Vegas, NV.
2. S. Broom, (2006) "Voip quality assessment: Taking account of the edge-device," Audio, Speech Lang. Processing (Special Issue on Objective Quality Assessment of Speech and Audio), vol. 14, no. 6, pp. 1977–1983, Nov.
3. N. Ct, V. Koehl, V. Gautier-Turbin, A. Raake, and S. Mller, (2010) "An intrusive super-wideband speech quality model: Dial," Speech Communication Association (Interspeech10), Makuhari, Japan.
4. W. Zhang, Y. Chang, Y. Liu, and L. Xiao, (2013) "A new method of objective speech quality assessment in communication system," Journal of Multimedia, Vol. 8, No. 3, June 2013, Academy Publisher.
5. Q. Li, Y. Fang, W. Lin, and D. Thalmann, (2014) "Non-intrusive quality assessment for enhanced speech signals based on spectro-temporal features," International Conference on Multimedia and Expo Workshops (ICMEW).
6. P. Bauer, C. Guillaumea, W. Tirry, and T. Fingsheidt, (2014) "On speech quality assessment of artificial bandwidth extension,"Conference on Acoustics, Speech and Signal Processing (ICASSP).
7. P. Reichl, S. Egger, R. Schatz, and A. DAlconzo, (2010) "The logarithmic nature of qoe and the role of the weber-fechner law in qoe assessment,"International Conference on Communication (ICC).
8. V. Emiya, E. Vincent, N. Harlander, and V. Hohmann, (2011) "Subjective and objective quality assessment of audio source separation,"Audio, Speech and Language Processing Vol. 19, No. 7.

9. T. Falk, C. Zheng, and W.-Y. Chan, (2010) "A non-intrusive quality and intelligibility measure of reverberant and dereverberated speech," Audio, Speech Lang. Processing (Special Issue on Processing Reverberant Speech: Methodologies and Applications), vol. 18, no. 7, pp. 1766–1774.

10. M. Gueguin, R. LeBouquin, V. Gautier-Turbin, G. Faucon, and V. Barriac, (2008) "On the evaluation of the conversational speech quality in telecommunications,"EURASIP J. Adv. Signal Processing, vol. 2008, Article ID 185248.

11. B. Weiss, S. Moller, A. Raake, J. Berger, and R. Ullmann, (2008) "Modeling conversational quality for time-varying transmission characteristics,"Acta Acoust. United Acoust., vol. 95, no. 6, pp. 1140–1151.

12. M. Sauter, (2010) "From gsm to lte: an introduction to mobile networks and mobile broadband," in John Wiley & Sons.

13. W. S. Cleveland and S. J. Devlin, (1988) "Locally Weighted Regression: An Approach to Regression Analysis by Local Fitting," Journal of the American Statistical Association, vol. 83, no. 403, pp. 596–610.

Chapter 6
Profiling Wireless Resource Usage for Mobile Apps

This chapter is devoted to the other side of mobile phone usage, namely the mobile apps. With the spread of LTE technology, network traffic induced by apps brings huge overhead to the mobile network. Profiling wireless resource usage is fundamental for getting better knowledge and identification of the network traffic. It is important to anticipate potential overloading of the bandwidth consumption. Also, it allows improving resource allocation for better quality of services.

On the one hand, the main studies in this domain are related to bandwidth profiling of each user, without specifically taking into account the diversity of applications of each smartphone. On the other hand, there is also a lot of work about profiling app's usage of smartphone resource at the phone side. In this chapter, a combination of those two aspects is combined to provide a profiling of wireless resource usage for mobile apps.

The main difficulties for this task are to leverage data and to connect the app behavior with network resource usage. For data collection, an app called AppWiR handles app traffic data and sends them to a server. For explore influence of each app on the network, a two-layer mapping model is developed. This model includes a feature selection algorithm based on random forests, combined with an SW-LOESS algorithm for quantitative mapping. Resulting profiling mechanism is effective and can accurately estimate and predict resource usages for mobiles apps.

6.1 Motivation and System Overview

6.1.1 Background and Challenges

Data and signaling storms intensively consume cellular resources, and significantly reduces the quality of service in cellular network. It attracts serious attention from network owners for understanding mobile apps' usage of network

© Springer International Publishing AG, part of Springer Nature 2018   83
Y. Ouyang et al., *Mining Over Air: Wireless Communication Networks Analytics*,
https://doi.org/10.1007/978-3-319-92312-3_6

resource, which is especially critical for resource control, quality of service improvement, and usage pricing.

However, there is little research considering the impact of mobile apps when analyzing the network resource. Almost all existing research focuses on profiling apps' usage of smartphone resources, and cannot be applied for profiling apps' usage of mobile resource. This is because they either focus solely on resource usage analysis at the phone side or do not consider the impact of mobile apps when analyzing network resource usage. The network resources are not directly influenced by apps but hundreds of wireless conditions, such as traffic volumes and signaling intensity, etc. They are shared and simultaneously consumed by thousands of network functions in different layers, such as handovers and app message transmission. The usage of lower-layer network resources by upper-layer network functions are mixed, due to the sharing nature and stacked network architecture. The app usage is harder to quantify or differentiate if the studied network functions are in the several layers. Moreover, due to the lack of relation between app behavior and network resource, it is hard to find the root cause that leads to the change of resource consumption, and thus cannot pinpoint the possible issue or forecast the usage. It is also difficult to clearly separate the resource usage of one app from the others due to the huge number of coexisting apps and their simultaneous impact to the network. For each particular mobile app, it is used in different time and locations with different network conditions, and thus its behaviors, network characteristics, ultimately, and the resource usage varies frequently. Hence, a scalable and light-load data collection method is needed.

6.1.2 Mobile Resource Management

Current research work related to mobile resource management can be categorized into two categories: Profiling app's usage of smartphone resource [1–4] and Network resource management and optimization [5–13]. The former stays at the device side and profiles the smartphone resource usage for mobile apps. The latter analyzes how user activities and mobility patterns affect mobile network resource allocations. Neither profiles mobile app's network resource usage at the network side.

Regarding collecting data from mobile apps, the existing resource profiling mechanisms rely on two methods. The first one is to collect data with specialized equipment, such as test phones and log recorder, etc., operated by domain experts. This involves huge efforts from human labor and hardware investments, and thus is not feasible for large-scale profiling. Another method is performing Deep Packet Inspection (DPI) at cell level to collect app-specific indicators. However, as DPI significantly consumes network resources to collect enough training data, it needs to be operated for long time and thus drastically downgrades regular network services. As far as we are aware, the system in this chapter is the first study to profile network resource consumption for mobile apps through user crowdsourcing and analytics. The proposed scheme builds on causal relationship between apps, network signals/

traffics, and network-side hardware usages, which greatly helps domain experts for resource usage diagnosis.

6.1.3 System Overview

We observe that mobile apps impact the networks by generating various types of data and signaling traffic. Such traffics directly consume the backend network resources. Based on this observation, the network traffics can be utilized as bridge to link the mobile app behaviors and network resource usages, and further to explore the causal relationship between them.

In this chapter, we introduce AppWiR, a crowdsourcing-based wireless analytical system, which collects app behavior information from smartphones and analyzes these information with network traffics and resources to build a mapping model. AppWiR consists of a crowdsourcing tool installed in smartphones to collect mobile app behavior for model training, and a set of analytical algorithms located in the server, which mine the collected data to model the relationship between mobile app behaviors and their network resource usages. Figure 6.1 illustrates the architecture of the AppWiR system.

We leverage the existing numerous mobile users, and implement the AppWiR crowdsourcing App, which is installed in Android smartphones and runs silently in background, to record behavior related indicators, such as app turn on/off, heart beats, sending/receiving messages, hand-over, and data traffics, etc., for each mobile app installed on the phone. We also collect the cell-level network traffic and resource usage data. To ensure collected cell-level data is consistent to the crowdsourced app data, we reserve testing cells that only allow app traffics from phones with the AppWiR tool. In this way we ensure that in the collected data, the network resource consumption completely reflects the activity of the crowdsourced mobile apps. A two-layer mapping model is built to establish the relationship between app behavior

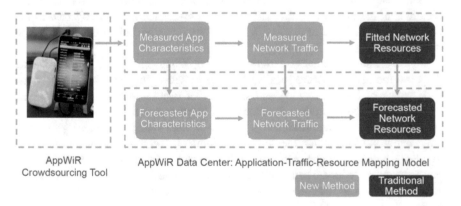

Fig. 6.1 The architecture of the AppWiR system

and network resource usage using the data we collected. One layer is created between mobile app and the network traffics, and the other one is built between network traffics and underlying resource usages. For each layer, Proximity Matrix assisted Feature Selection (PMFS) algorithm is utilized to select relevant indicators highly correlated to the state change of the layer, by applying Random Forest decision trees. After that, a Sliding-Window based Local Weight Scatterplot Smoothing (SW-LOESS) algorithm is developed to regress the selected indicators and establish the mapping in this layer. Such mapping mechanism is applied to both layers.

With the two-layer mapping, the change of app behaviors can successfully reflect the change of network traffics, and further propagates to reflect the change of resource usages. Therefore, it is able to use the indicators of each mobile app to model and estimate its network resources consumption. We validate the established model, represent the discovered knowledge in time series, and further analyze its temporal characteristics to predict future the network resource usages.

6.2   AppWir Crowdsourcing Tool

A distributed AppWiR crowdsourcing App is developed and installed in Android phones to record hundreds of network and behavior indicators for selected apps running in the phone. The AppWiR can be dynamically configured to cooperate with different test environments. After configuration, the App runs in background and logs the pre-specified app indicator without disrupting any cellphone usages.

The collected app data is compressed and temporarily stored in the phone, where each record is labeled using the name of the corresponding app. After a certain time, the App automatically packs the recorded indicator logs and sends them to the data center. Upon receiving the logs, for each app of interest, all the collected records of behavior indicators from different phones are merged. The AppWiR App runs 24/7 in each installed phones until it collects enough data for training the mapping model. Meanwhile, the network traffic and resource usage data are also collected at the corresponding cell in a centralized fashion since it is trivial and does not needs to differentiate apps.

Figure 6.2 shows the user interface of the AppWiR App. To help domain experts diagnose the network issues locally, the App provides functions for configuration and display of log information and summary of results, such as app states, data throughput, connection types, durations, and location traces, etc. Such information can be displayed in real-time and shown in several types. Once enough data is collected at the data center, the AppWiR mining algorithm is executed to train the two-layer mapping model, and the results are stored in the server. In the testing phase, to apply the model and profile the resource usage for apps, we turn on DPI for an hour to collect app data, since it does not need as much data as used in the training phase. The collection of DPI data is executed once per month, which does not bring

Fig. 6.2 User interface of the AppWiR tool

much overhead to the network. After collecting the data to the server, the trained model is applied to mine the collected app behavior data to estimate the historical network resource usage for each interested app, and forecast its future usages.

6.3 AppWir Mining Algorithm

6.3.1 Selection of Network Indicators

Proximity Matrix assisted Feature Selection (PMFS), which leverages Random-Forest decision trees to score the importance of each indicator according to its similarity distance, is used in the system to select relevant indicators for building the two-layer mapping model.

Each indicator of each record is given a label according to its measured value after the data collection. With labels, we apply Random Forest classifier to build decision trees and classify the collected data into different classes. While building the trees, a two-dimensional similarity matrix named Proximity Matrix is formed, where each entry records the similarity distance between each pair of the indicators. We use the designed Proximity Matrix to measure the similarity between clusters, and apply such knowledge to score the capability of each indicator in separating data into different classes. Only indicators with high scores are selected as features. More specifically, the Proximity Matrix is developed and updated during the growth of the decision trees in Random Forest. Given a training dataset with n indicators, initially, the Proximity Matrix P is a n × n matrix with all zeros. When the trees grow, we examine each node of the trees. At a tree node, if a pair of indicators, fi and fj, both appear in this node, then we update the matrix entry P_{ij} by adding 1 (i.e.,$P_{ij} = P_{ij} + 1$). This examination process is repeated until all the decision trees are fully grown in the forest. After that, we normalize the value in each entry of the matrix and obtain the developed Proximity Matrix, in which each entry represents the similarity of the corresponding pair of indicators.

We compute the importance score for each indicator based on the Proximity Matrix. Assuming that a training dataset has n indicators, and is classified into c classes, we compute the ratio of intra-class similarity, P_{intra}, and inter-class similarity, Pinter, as:

$$R = P_{intra} / P_{inter} \tag{6.1}$$

where $P_{intra} = \sum_{i,j=1}^{n} P_{i,j}$ (if i = j) and $P_{inter} = \sum_{i,j=1}^{n} P_{i,j}$ (if i $\neq$ j). For each indicator fi, we replace its value with random noise and obtain a new dataset to determine its importance. We feed the new dataset to Random Forest classifier and obtain a new Proximity Matrix P_i along with its corresponding similarity ratio R_i. We compute the difference between the new ratio and the original one, $R'_i = R - R_i$, and apply this process for all the indicator and obtain the difference of similarity ratio for each of them. Finally, we calculate the importance score IS_i for each indicator f_i by normalizing its difference of similarity ratio as $IS_i = R'_i / S$, where S is the standard deviation of the difference of all the indicators $\{R'_1, R'_2, \ldots, R'_n\}$. In this model, the higher an indicator's importance score is, the more important and relevant the indicator is to the classifier. Therefore, we can select indicators with high importance score considering they can be used for indicating the change of the data (i.e., the change of network resource). It is worth to note that, there are thousands of indicators in cellular networks, and it may be time consuming to score the relevance for all of them. To reduce the search space, we use domain knowledge to preselect a list of indicator candidates, and only search among the selected candidates instead of all the indicators.

According to 3GPP TR 36.942, we first categorize TCP Power into 4 categories [0dbm, 10dbm], (10dbm,20dbm], (20dbm,30dbm], and (30dbm,43dbm], and place a label for each of them. We apply Random Forest classifier and train 1500 trees to derive the Proximity Matrix for TCP Power and get the importance score. After normalization, Table 6.1 shows the top 11 traffic indicators that are highly relevant to the TCP Power.

Table 6.1 Selected traffic indicators

Categories of traffic indicator	Traffic indicator	Important score
User plane indicators	DL.Cell.PRB.Used.Average	0.8735
	DL.Cell.Simultaneous.Users.Average	0.8454
	DL.Cell.PDCP.Throughput	0.8253
	Cell.RRC.Connected.Users.Average	0.8192
Signaling plane indicators	Cell.RRC.Connection.Req	0.7960
	Cell.eRAB.Setup.Req	0.7807
	Cell.Paging.UUInterface.Number	0.7402
	Cell.PDCCH.OFDM.Symbol.Number	0.7396
	Cell.PDCCH.OFDM.CCE.Number	0.7308
Mobility indicators	Cell.Intra+IntereNB.Handover.Out	0.6377
	Cell.Intra+IntereNB.Handover.In	0.6169

Table 6.2 Selected APP behavior indicators

APP behavior indicators	Important score
DL.TrafficVolumn.Bytes.PerApp	0.8690
DL.MeanHoldingTime.PerSession.PerApp	0.8529
Sessions.PerUser.PerApp	0.8181
ActiveSessions.PerApp	0.8116
Registered.Users.PerApp	0.8012
DL.ActiveUsers.PerApp	0.7921
Throughput.PerSession.PerApp	0.7408
DL.PacketCall.Frequency.PerApp	0.7134
UL.ActiveUsers.PerApp	0.7103
DL.Bytes.PerPacketCall.PerApp	0.6945
DL.Packets.PerPacketCall.PerApp	0.6733
PacketFreq.PerPacketCall.PerApp	0.6402
DL.PacketCalls.PerSession.PerApp	0.6307

In Table 6.1, the selected traffic indicators are roughly categorized to the following three categories. The mobility indicators are Intra/Inter eNodeB handovers incoming direction and outgoing direction respectively.

The selected indicators and their corresponding categories are under our expectation since in reality these three categories are the major facts that intensively consume the wireless resource. Similarly, we apply PMFS to select app behavior indicators based on the selected traffic indicators. Table 6.2 lists the top app indicators that highly impact traffic indicators.

6.3.2 LOESS Method

Sliding-Window based Local Weight Scatterplot Smoothing (SW-LOESS) is developed in the regression, which improves LOESS by dynamically computing a proper window size during the regression. We treat the selected indicators as features, pack the value of each feature into different bins, and dynamically adjust window size for each local set based on the distribution of each bin. Practically, such bins can be configured by domain experts based on their experience. After the bin configuration, given a feature with n points and k bins, each with equivalent length L = n/k, we set an initial window size to n/100, and plot the scatterplot for all its measured values in an ascending order. Let f(x) denotes the scatterplot function. First, for each bin bin_j we compute its distribution density by integrating the value of the scatterplot function in its range as follows:

$$F_j = \int_{f^{-1}(L*j)}^{f^{-1}(L*j+L)} f(x)dx \ , (j = 0, \ldots, k - 1) \tag{6.2}$$

$F = \{F_0, F_2, \ldots, F_{k-1}\}$ is sorted in ascending order. BF_{min} represent the bin with minimum value in F, BF_{med} represents the bin with median value, and BF_{max} is the bin with maximum value, we dynamically calculate the window size by sorting the results as:

$$win\_size = \begin{cases} \dfrac{0.5(1 + 1/i).B}{100}.N & , B = 0, \ldots, i \\[3mm] \dfrac{1 + (B - i)}{100}.N & , B = i + 1, i + 2, \ldots, k \end{cases} \tag{6.3}$$

The dynamically calculated window size is utilized to perform LOESS regression to the selected features in each of the two layers. After the regression, we successfully obtain the two-layer mapping, where we can use the indicators of mobile app behaviors to model network traffics, and further use the network traffics to model network resources.

6.3.3 Usage Prediction with Time Series Analysis

In this section, the prediction model is leveraged to design a temporal mining algorithm to predict the future app behavior that can be used for forecasting the future network resource usages. In the AppWiR system, the application characteristics indicators are crowd-sourced from mobile users and averaged at each cell. For a behavior indicator, its time series in a given time interval is directly measured and is a mixture of various characteristics, such as trending, seasonal change, sudden change, random fluctuation, and even noise. To clearly understand how each indicator evolves along time, we design an algorithm to decompose the measured time series into four components: Trend component T(t), Seasonality component S(t), Burst component B(t), and Random noise component R(t).

T(t) represents a long term change of app behaviors, such as user behaviors, rate plan policy, or user numbers, and reflect the change with large granularity(e.g. per week). S(t) denotes periodical changes, such as repeated daily changes of application traffic (busy hour/non-busy hour). B(t) indicates a significant change from normal trend which is caused by external known or unknown factors, and R(t) contains unpredictable fluctuations and measurement noises. After component extraction, we predict each of the extracted components and add the predicted components together to form the final prediction result.

6.4 Implementation and Trial

6.4.1 Data Collection and Study

The AppWiR trial is conducted in LTE network of a leading carrier under different network conditions, to test accuracy, reliability, and predictability of the AppWiR system and evaluate its diagnosability for identifying most relevant app behaviors to wireless resource.

AppWiR is first deployed to collect data from mobile users and build the two-layer mapping model. This process lasts 2 months from January to February 2014. The AppWiR App receives downloads from 50 smartphones with Android 4.2+ System that is compatible with all major applications such as Facebook, YouTube, WeChat, What's App, GoogleMap, etc. The AppWiR App logs all the needed application behavior indicators and generates and periodically uploads the testing logs to the data center. To make sure the collected app behavior data is consistent to the network usage data, we deployed four testing cells adjacent to each other, in which an IMEI list is configured such that only those 50 smartphones are allowed to access the testing cells. Any other devices that intend to access or handover into the testing cells will be blocked. In this way, we ensure the application data generated through the 50 smartphones is completely synchronized and in line with the traffic statistics logs generated in the testing cells.

To evaluate the trained model, we use DPI to collect app behavior data and test the profiling and prediction accuracy of the model. This step lasts for 7 months, from February to July 2014, which is longer than the first step to obtain the temporal trend and seasonality of the data. In this step, to test our model in regular cells, we do not use test cells. Instead, we enable DPI 30 min per week in regular cells to collect data. The measured DPI data consists of behavior indicators of various apps and is aligned with the granularity of traffic statistics logs. We select Downlink Cell Transmitted Power (TCP Power) as the network resource indicator of interest, because it is the most critical resource that support major functionality of the network. The trial will then analyze how mobile apps consume the TCP Power.

Two kinds of data sets are collected during the trial. The first data set is application logs crowd-sourced by the AppWiR App and the corresponding traffic and resource usage statistics from the testing cells. The second data set is the DPI logs. In total, we deep inspect the network for 207 busy hours and collect the data. We exclude 10 h of data due to either incomplete logs or parsing failure, and obtain 197 valid busy hour measurements, which is used for testing the designed model and validating the forecasting algorithms.

6.4.2   Results and Accuracy

80% of the entire data set are utilized as training set, and the rest of 20% are used as test set in the evaluation. We compare the indicator value computed by the AppWiR model with the ground-truth measured value, and use the Mean Absolute Percentage Error (MAPE) to calculate the error for the model through the Eq. 6.4.

$$e = \frac{1}{n} \sum_{i=1}^{n} \left| \frac{S_i^{measure} - S_i^{est}}{S_i^{measure}} \right| \qquad (6.4)$$

where Simeasure and Siest are the measured and estimated indicators corresponding to the i-th app respectively. The MAPE of the 11 selected traffic indicators is displayed in Fig. 6.3. From Fig. 6.3, it is clear that the testing MAPE of all the traffic indicators except mobility related indicators are less than 0.25, and the training MAPE is even lower. The higher MAPE on mobility indicators is because our model is trained using data in the four testing cells, while is tested in many widely distributed regular cells with DPI data. The testing cells are near to each other and thus cannot capture enough mobility behaviors as shown in regular cells, and thus the MAPE of mobility related indicators are higher than the others. However, as the importance score of mobility indicators are low, less than 0.65 as shown in Table 6.1, their MAPE does not impact the overall accuracy of the model much.

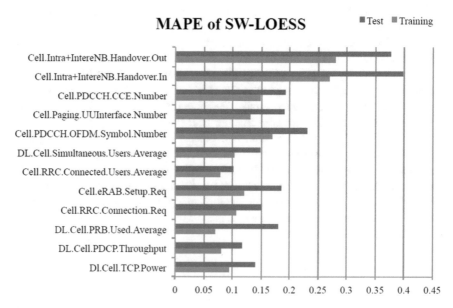

Fig. 6.3 MAPE of the estimated traffic indicators

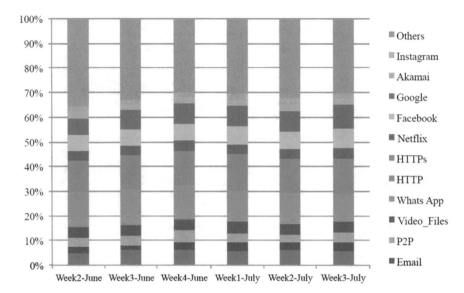

Fig. 6.4 Percentage of network resource usages by different apps

Hundreds of mobile apps are tested. Figure 6.4 shows the percentage of resource usages (TCP Power) by major apps. From this figure we can see that HTTP/HTTPS, i.e., web browsing apps, ranks highest in resource consumption because web browsing is always the top one usage for smartphones. Streaming applications, such as P2P, Netflix, and video-file related apps, also consume resource heavily. In addition to these two categories, signal-heavy applications such as Facebook and WhatsApp together consume considerable power resource, due to their large amount of users. Such profile enables mobile operators to understand how each mobile application consumes wireless network resource, and significantly help them to manage and price the usages.

The time series prediction algorithm is applied to predict the app behavior indicators. Figure 6.5 shows the prediction result for two typical app indicators: Downlink Active Users and Uplink Active Users. From this figure we can see that the training MAPE of two indicators is 7.47% and 8.93%, respectively, while the prediction (test) MAPE goes a bit up to 12.54% and 13.39%. The MAPE difference between training and prediction set is about 5%, which shows the prediction model is reliable and robust. We also apply such prediction algorithm to other app indicators, and their training MAPE is between 7.47% and 18.34%, while prediction MAPE is between 12.54% and 25.78%. Overall, majority of the indicator's prediction MAPE is lower than 15%. The highest one is from indicator DL.PacketCalls.PerSession. PerApp, which is caused by unstable application portfolio in the cell in different sampling hours. For example, in a cell, for a certain time, majority of its traffic is from YouTube, while right after that, all traffics switch to Instant Messages. Such a drastic change of application portfolio causes large variance for this indicator, which makes it difficult to capture its long term trend, midterm seasonality, and short term

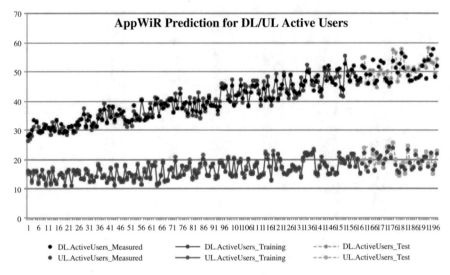

Fig. 6.5 Prediction of app behavior indicators

spikes. On the other hand, such observation also indicate the reason that this indicator has the lowest importance score in the AppWiR mapping model as shown in Table 6.2.

References

1. F. Qian, Z. Wang, A. Gerber, Z. Mao, S. Sen, and O. Spatscheck, (2011) "Profiling resource usage for mobile applications: a cross-layer approach," MobiSys.
2. D. Narayanan, J. Flinn, and M. Satyanarayanan, (2000) "Using history to improve mobile application adaptation," Workshop on Mobile Computing Systems and Applications.
3. L. Ravindranath, J. Padhye, S. Agarwal, R. Mahajan, I. Obermiller, and S. Shayandeh, (2012)"Appinsight: Mobile app performance monitoring in the wild," USENIX OSDI.
4. Y. Xu, M. Lin, H. Lu, G. Cardone, N. Lane, Z. Chen, A. Campbell, and T. Choudhury, (2013) "Preference, context and communities: a multi-faceted approach to predicting smartphone app usage patterns," Symposium on Wearable Computers.
5. T. Ng and W. Yu, (2007) "Joint optimization of relay strategies and resource allocations in cooperative cellular networks," Journal on Selected Areas in Communications, vol. 25, 2.
6. C.-H. Yu, K. Doppler, C. Ribeiro, and O. Tirkkonen, (2011) "Resource sharing optimization for device-to-device communication underlaying cellular networks," Wireless Communications, vol. 10, 8.
7. K. Zheng, F. Hu, W. Wang, W. Xiang, and M. Dohler, (2012) "Radio resource allocation in lte-advanced cellular networks with m2m communications," Communications Magazine, vol. 50, 7.
8. W.-H. Park and S. Bahk, (2009) "Resource management policies for fixed relays in cellular networks," Computer Communications, vol. 32, 4, Pages 703711.

9. K. Doppler, S. Redana, M. Wdczak, P. Rost, and R. Wichman, (2009) "Dynamic resource assignment and cooperative relaying in cellular networks: concept and performance assessment," Journal on Wireless Communications and Networking, vol. 2009, no. 24.
10. Abdelhadi and C. Clancy, (2014) "Context-aware resource allocation in cellular networks," in arXiv:1406.1910 [cs.NI].
11. H. Shajaiah, A. Abdelhadi, and C. Clancy, (2014) "Multi-application resource allocation with users discrimination in cellular networks," in arXiv:1406.1818 [cs.NI].
12. D. Fooladivanda and C. Rosenberg, (2013) "Joint resource allocation and user association for heterogeneous wireless cellular networks," Wireless Communications, vol. 12, 1.
13. Q. Ye, B. Rong, Y. Chen, M. Al-Shalash, C. Caramanis, and J. G. Andrews, (2013) "User association for load balancing in heterogeneous cellular networks," Wireless Communications, vol. 12, 6.

Chapter 7
Outlier Identification in Telecom Data

This chapter introduces technology to detect anomalies caused by technical equipment problems or fraudulent intrusion in telecommunication networks. The anomaly detection technology extracts information from network raw data and utilizes machine learning algorithms to alert network managers when an anomaly occurs.

Since data collected in telecom networks contains values for different features as well as time stamps, such data can be modeled and processed to seek and detect anomalies using unsupervised algorithms in machine learning. Algorithms leverage unlabeled data and assume that information about which data elements are anomalies is unknown. The algorithm does not directly detect anomalies but instead separate and distinguish data structures and patterns in order to group data from which "zones of anomalies" are deduced. Time stamps information is commonly collected when data are generated but is not widely used in classic anomaly detection processes. By adding time-stamp attributes in an evaluation model, it can allow us to discover periodic behaviors.

In this chapter, we introduce an unsupervised model to detect anomalies, focuses on algorithms combining both values and date (time stamps) and discuss two new models to this end.

7.1 Models

Unsupervised models are utilized to detect anomalies in telecommunication network. Specifically, it focuses on algorithms combining both feature values and dates (time stamps) and introduce two new models to this end. The first one is the time-dependent Gaussian Mixture Model (time-GMM), which is a time dependent extension of GMM [1] which works by considering each period of time independently. The second one is Gaussian Probabilistic Latent Semantic Analysis (GPLSA), derived from Probabilistic Latent Semantic Analysis (PLSA) [2], which combines values and dates processing together in a unique machine learning algorithm. This

© Springer International Publishing AG, part of Springer Nature 2018   97
Y. Ouyang et al., *Mining Over Air: Wireless Communication Networks Analytics*,
https://doi.org/10.1007/978-3-319-92312-3_7

latter algorithm is well known in text-mining and recommender systems areas but has been rarely used in other domains such as anomaly detection.

In this chapter, these two algorithms are implemented with R [3]. Their abilities are tested to find anomalies and adapt new patterns on both sample and traffic data. Five anomaly detection models are presented in this chapter: Gaussian, time-dependent Gaussian, GMM, time-dependent GMM, and GPLSA. The following notations are defined and used for all models.

W is a traffic data set. This set contains N values indexed with i. N is usually large, that is, from one thousand to one hundred million. Each value is a vector of R^p, where p is the number of features. Furthermore, each feature is assumed to be continuous. D is the time-stamp set of classes. This set also contains N values. Since we are expecting a daily cycle, each value d_i corresponds to each hour of the day, consequently standing in $\{1,\ldots, 24\}$. $X = (W, D)$ are observed data. For clustering methods, it is assumed that each value is related to a fixed cluster, named Z. It is a "Latent" set, since it is initially unknown. The number of clusters K is known.

For each model, the aim is to estimate parameters with maximum likelihood. When the direct calculation is intractable, an EM algorithm is used to find a local optimum of the likelihood. A usual hypothesis of independence is added, which is needed to compute the likelihood of the product over the set. The set of triples (W_i, Z_i, D_i) is an independent vector over the row i. Note that if the model does not consider D or Z, the set will be removed.

7.1.1 Gaussian Model

In the Gaussian model, the whole data set is assumed to come from a variable that follows a Gaussian distribution. Consequently, each part of the day has a similar behavior and there are no clusters. Each variable W_i follows Gaussian distribution with mean and variance m, $\sum$. Here, m is a p-vector and $\sum$ is a variance-covariance matrix of size p. they are both independent of i. Parameters are easily estimated with empirical mean and variance.

7.1.2 Time-Dependent Gaussian Model

A time component is added in this model, as opposed to the Gaussian model, which does not include a time component. Each time of the day is considered independently, following a particular Gaussian distribution. This allows us to take dependence of time into account. For each s in $\{1,\ldots,24\}$, each conditional variable W_i such that $D_i = s$ follows a Gaussian distribution with mean and variance m^s, $\sum^s$. As for the Gaussian model, parameters are estimated with empirical mean and variance for each class of dates.

7.1.3   Gaussian Mixture Model

Compared to the Gaussian model, in the GMM, data is assumed to come from a mixture of Gaussian distributions rather than one single Gaussian distribution. The number of clusters K is fixed in advance. In this model, each record belongs to a cluster $Z_i = k$ in $\{1,\ldots,K\}$ with probability a_k. Each variable $(W_i \mid Z_i = k)$ follows a Gaussian distribution with mean and variance m_k, Σ_k. In this model, each record belongs to an unknown cluster. The task is to estimate both probability for each cluster and the parameters of each Gaussian distribution. To solve this problem, the following decomposition is done:

$$p(W_i) = \sum_k p(W_i \mid Z_i = k)p(Z_i = k) \qquad (7.1)$$

The parameters can be successively updated with an EM algorithm.

7.1.4   Time-Dependent Gaussian Mixture Model

The time-dependent GMM is designed including both clustering and time-dependence. The EM algorithm is used to estimate parameters. For each s in $\{1,\ldots,24\}$, each record such that $D_i = s$ belongs to a cluster $Z_i = k$ in $\{1,\ldots,K\}$ with probability $a_{k,s}$. For each s in $\{1,\ldots,24\}$, each variable $(W_i \mid Z_i = k)$ such that $D_i = s$ follows a Gaussian distribution with mean and variance m_k^s, $\sum_k^s$.

7.1.5   Gaussian Probabilistic Latent Semantic Model

The GPLSA model is based on the classic GMM but introduces a novel link between data values and time stamps. In time-GMM, the different classes of dates are considered independently, whereas GPLSA introduces dependence between latent clusters and time stamps but only within those two variables. That is, in knowing latent cluster Z, it assumes that there is no more dependence on time. This assumption allows making the problem computationally tractable. Explicitly, the following occurs:

For each s in $\{1,\ldots,24\}$, each reach such that $D_i = s$ belongs to a cluster $Z_i = k$ in $\{1,\ldots,24\}$ with probability $a_{k,s}$. Each variable $(W_i \mid Z_i = k)$ follows a Gaussian distribution with mean and variance m_k, Σ_k. For all i, $P(W_i \mid D_i,Z_i) = P(W_i \mid Z_i)$. To solve this problem, the following decomposition is done:

$$p(W_i|D_i = s) = \sum_k p(W_i|Z_i = k)p(Z_i = k|D_i = s) \qquad (7.2)$$

The EM algorithm can be adapted in this case to iteratively increase the likelihood and estimate parameters in order to obtain exact update formulas. It let $(\cdot \mid m, \Sigma)$ equal the density of a Gaussian with parameters m and Σ. Also, E_s is defined as the set of indexes i, where $d_i = s$. The following algorithm describes the steps to get final parameters:

Step 1: At time t $= 1$, let some initial parameters $m_k^{(t-1)}$, $\sum_k^{(t-1)}$, and $a_{k,s}^{(t-1)}$ for all k and s.

Step 2: For all k, i, compute the probability $Z_i = k$ knowing $W_i = \varpi_i$, $D_i = d_i$, and parameters:

$$T_{k,i}^{(t)} = \frac{f(\varpi_i|m_k^{(t-1)}, \sum_k^{(t-1)})a_{k,d_i}^{(t-1)}}{\sum_{l=1}^{K} f(\varpi_i|m_l^{(t-1)}, \sum_l^{(t-1)})a_{l,d_i}^{(t-1)}} \qquad (7.3)$$

Step 3: For all K, s, compute the following result:

$$S_{k,s}^{(t)} = \sum_{j=1}^{\#E_s} T_{k,E_s(j)}^{(t)}$$

Step 4: For all k, s, update $a_{k,s}$ with:

$$a_{k,s}^{(t)} = \frac{S_{k,s}^{(t)}}{\sum_{l=1}^{K} S_{l,s}^{(t)}}$$

Step 5: For all k, update the means with:

$$m_k^{(t)} = \frac{\sum_{i=1}^{K} w_i T_{k,s}^{(t)}}{\sum_{l=1}^{N} T_{k,i}^{(t)}}$$

Step 6: For all k, update the covariance matrix with:

$$\sum_k^{(t)} = \frac{\sum_{i=1}^{K} (\varpi_i - \varpi_k)'(\varpi_i - \varpi_k)T_{k,i}^{(t)}}{\sum_{l=1}^{N} T_{k,i}^{(t)}}$$

Step 7: Let $t = t + 1$ and repeat Step 2–7 until convergence at a date T, at that date, parameters are estimated.

Step 8: For each i, the chosen cluster is k maximizing $T_{k,i}^{(t)}$

Step 9: For each i, the likelihood of this point for the estimated parameters is:

$$P(d_i) \sum_{l=1}^{K} f(\omega_i | m_l^{(T)}, \sum_l^{(T)}) a_{l,d_i}^{(T)}$$

7.2 Comparison of Models

All five models defined in the above section are implemented with R into a framework that is able to perform computations and to show clustering and anomaly identification plots. In this section, a framework is designed to compare abilities to detect anomalies and check robustness of the methods. The sample set is built to highlight the difference of behaviors between models in a simple and understandable context. Consequently, only one sample feature is considered in addition to time-stamp dates. In this set, time-GMM and GPLSA are able to detect anomalies within the set, and those methods are then potential candidates for anomaly detection in a time-dependent context [4, 5–7]. Furthermore, it shows that GPLSA is more robust and allows a higher interpretation level of resulting clusters.

7.2.1 Sample Definition

The sample is built by superposing the three following random sets.

$$t \rightarrow \cos\left(\frac{2\pi t}{T}\right) + \varepsilon$$

$$t \rightarrow \cos\left(\pi + \frac{2\pi t}{T}\right) + \varepsilon$$

$$t \rightarrow -2.5 + \varepsilon$$

where ε is independent random variables for each t sampled according to the continuous uniform distribution on $[0, 1]$ and where T has a daily period. The range of the two first functions is 24 h, whereas the third one is only defined from 0:00 to 15:00. Three anomalies are added on this set, defined, respectively, at 6:00, 12:00, and 18:00 with values -1.25, 0.5, and 1.65. The resulting set is shown in Fig. 7.1.

7.2.2 Anomaly Identification

All five models are trained and the likelihood of each point is computed for each model. Since we expect 3 anomalies to be found in this sample set, the 3 lowest likelihood values are defined as anomalies for each model. For the clustering process, the chosen number of clusters is $K = 5$. The results are shown in Fig. 7.1. In (a), the whole data set is modeled as one Gaussian distribution and we found no expected anomalies. In (b), each period is determined with a Gaussian distribution, and we only discovered the anomaly at 18:00. In (c), the whole set is clustered and we only discovered the anomaly at 6:00. Finally, in (d), the time GMM and GPLSA models are trained and the same results obtained: the 3 anomalies were successively detected. Thus, time-GMM and GPLSA are both able to detect expected anomalies contrary to other methods.

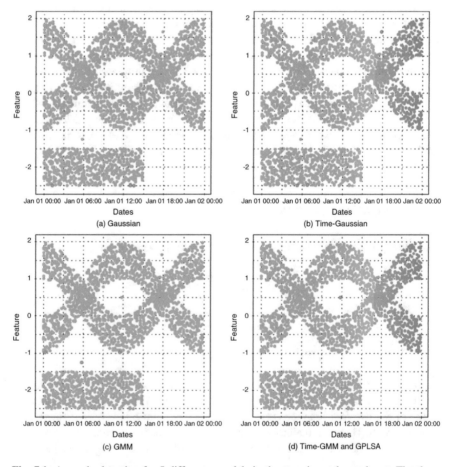

Fig. 7.1 Anomaly detection for 5 difference models in the experimental sample set. The three values with the lowest likelihood are circled in orange. Each color represents a different time-stamp class (only 1 class for (**a**) and (**c**); 24 classes for (**b**) and (**d**))

Table 7.1 Summary of anomaly detection results

Inserted anomalies (ID:60–71)	GMM	GPLSA
Red Zone	NA	66–71
Yellow Zone (w = 6)	NA	62–66

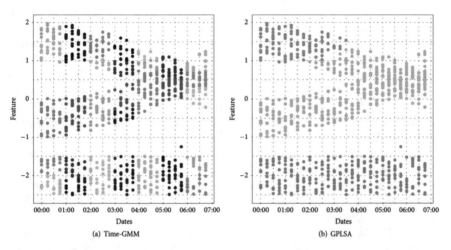

(a) Time-GMM   (b) GPLSA

Fig. 7.2 Identified clusters for 2 models in the experimental sample set between 0:00 and 7:00. In (**a**), each class of 1 h contains 5 clusters, and clusters are not related across hours. In (**b**), the whole set contains 5 clusters

7.2.3   Comparison Between Time-GMM and GPLSA

The same anomalies have been detected with time-GMM and GPLSA. However, they are detected differently. A summary of the comparison is showed in Table 7.1.

First, GPLSA evaluates time stamps and values at once; that is, all parameters are estimated at the same time. Consequently, consecutive dates can share similar clustering behaviors. With time-GMM, parameters are trained independently for each class of dates, and no relation exists between the clusters of different classes. Second, the number of clusters in each class is soft for GPLSA (i.e., it can be different to the specified number of clusters for some class of dates). This allows the model to automatically adapt the number of clusters depending on which cluster is needed in the model. In time-GMM, each class has a specified number of clusters. This is shown in Fig. 7.2, where the first 7 h are plotted in identified clusters for time-GMM (a) and GPLSA (b). Third, the model is trained with the whole data for GPLSA, whereas only a fraction of data is used for each time-GMM computation. If there is a limited number of data in a class of dates, this can cause a failure to correctly estimate time-GMM parameters. Fourth, the number of parameters needed for estimation is $(D + 2) \times K$ for GPLSA and $(3K - 1) \times D$ for time-GMM (with D number of classes and K number of clusters, and in dimension p = 1). Consequently, there are fewer parameters to estimate with GPLSA.

On the whole, GPLSA implies a better interpretation level (first and second points) of resulting clusters over time-GMM, combined with a higher robustness (third and fourth points).

7.3 Simulation and Discussion

In this section, anomaly detection is performed on real traffic network data. Based on
the comparison of models done in the above section, GPLSA is selected to deduce
anomalies and compare results with time-GMM.

Data were gathered from a Chinese mobile operator, which comprises a selection
of 24 traffic features collected for 3000 cells in the city of Wuxi, China. The features
are only related to cell sites and do not give information about specific users. They
represent, for example, the average number of users within a cell or the total data
traffic for the last quarter of hour. The algorithm is trained over 2 weeks, with one
value for each quarter of hour and for each cell. Rows of data containing missing
values are discarded. Only values and time stamps were taken into consideration for
computations, and the identification number of cells was discarded. Some features
only take nonnegative values and have a skewed behavior, and consequently, some
features are preprocessed by applying the logarithm. It expects that GPLSA can
manage this set, even though some properties of the model are not verified, such as
normality assumptions.

The GPLSA model is utilized for the feature corresponding to the "average
number of users within cell" and selected $K = 3$clusters. Anomalies are values
with the lowest resulting likelihood, computed to get (on average) 2 alerts and
8 warnings each day. Visual results are shown on Fig. 7.3. In (a), the three clusters
are identified, whereas, in (b), a different color is used for each class of dates. In (c),
the different log-likelihood values are shown. Finally, in (d), the estimation of the
probability α_k, to be in each cluster k knowing $D = s$ is plotted. Anomalies are shown
in (a), (b), and (c) and the extreme values related to each class of dates are correctly
detected. In (a) and (d), identified clusters are shown in three distinct colors. The
probability to be in each cluster varies across class as expected, with a lower
probability in the upper cluster during off-peak hours. Also, as shown in (a), the
upper cluster has a symmetric shape and the mean value is relatively similar across dates.

The results are obtained with time-GMM, using the same number of clusters
$K = 3$, and the same number of alerts and warnings each day. Results are shown on
Fig. 7.4. In (a), the three clusters are identified for each class D (between 1 and 24)
and in (b), the different log-likelihood values are shown. It is observed that time-
GMM correctly detects most of extreme values. Each class is related to a specific
likelihood function and has its own way to represent data. It shows that the cluster
extents related to the highest values have a similar width for all classes on Fig. 7.4a
($D = 1$ to 24). By comparing Fig. 7.4b with Fig. 7.3c, it is observed that a larger
"bump" (located in green during off-peak hours) for time-GMM. For these reasons,
and contrary to GPLSA, anomalies are overrepresented in some classes (e.g.,
3 warnings are detected for $D = 8$ for the first 2 days) whereas others do not contain
anomalies for this time period ($D = 6$). Those results endorse the higher level of
interpretation and robustness of GPLSA over time-GMM.

According to the results, GPLSA is able to detect anomalies in a time-dependent
context. Global outliers (e.g., on Fig. 7.3b at Apr. 15 16:00 in red) as well as context-

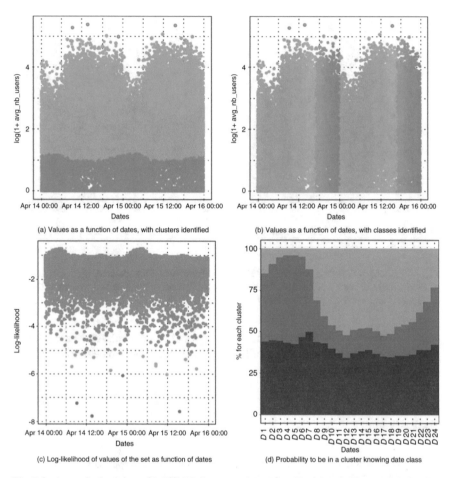

Fig. 7.3 Anomaly detection with GPLSA from experimental traffic data set. Plots are restricted to 2 days in (**a**), (**b**), and (**c**). Red and orange points are related to the lowest likelihood values obtained with an average of 2 red points and 8 orange points each day

dependent anomalies (e.g., at Apr. 15 5:00 in orange) are identified. Off-peak periods are taken into consideration and unusual values specific to those periods detected. Gaussian hypothesis on GPLSA is not really constraining. As shown in Fig. 7.3a, clusters are adaptable and try to fit Gaussian distributions. They are appropriate to represent the value distribution for each class of dates and cluster. Cluster adaptation is shown in Fig. 7.3d. The three clusters represent different level of values. The upper cluster represents higher values, which are more probable during peak periods. The lower cluster represents lower values, with a roughly constant probability. The third cluster in the middle is also useful to obtain a good anomaly detection behavior (results with $K = 2$ clusters are unable to correctly detect anomalies). About anomaly detection itself, a threshold indicating the number of alerts to be detected can be set. This method of detection is static and relatively

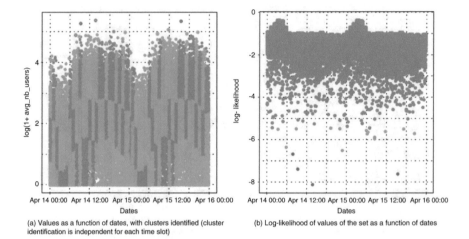

(a) Values as a function of dates, with clusters identified (cluster identification is independent for each time slot)

(b) Log-likelihood of values of the set as a function of dates

Fig. 7.4 Anomaly detection with time-GMM from experimental traffic data set. Plots are restricted to 2 days. Red and orange points are related to lowest likelihood values obtained with an average of 2 red points and 8 orange points each day

simple. Improving this method of detection is possible and straightforward through likelihood computations: inside a cell, an anomaly could be detected with a repetition of low likelihood scores.

References

1. G. J. McLachlan and K. E. Basford, "Mixture models. Inference and applications to clustering," in Statistics: Textbooks and Monographs, Dekker, New York, NY, USA, 1988.
2. T. Hofmann, "Probabilistic latent semantic analysis," in Proceedings of the 15th Conference on Uncertainty in Artificial Intelligence, Morgan Kaufmann, 1999.
3. R Core Team, R: A Language and Environment for Statistical Computing, R Foundation for Statistical Computing, Vienna, Austria, 2016, https://www.R-project.org/.
4. V. Chandola, A. Banerjee, and V. Kumar, "Anomaly detection: a survey," ACM Computing Surveys, vol. 41, no. 3, article 15, 2009.
5. P. Laskov, P. D'ussel, C. Sch"afer, and K. Rieck, "Learning intrusion detection: supervised or unsupervised?" in Image Analysis and Processing—ICIAP 2005: 13th International Conference, Cagliari, Italy, September 6–8, 2005. Proceedings, vol. 3617 of Lecture Notes in Computer Science, pp. 50–57, Springer, Berlin, Germany, 2005.
6. N. V. Chawla, N. Japkowicz, and A. Kotcz, "Editorial: special issue on learning from imbalanced data sets," ACM Sigkdd Explorations Newsletter, vol. 6, no. 1, pp. 1–6, 2004.
7. C. Phua, D. Alahakoon, and V. Lee, "Minority report in fraud detection: classification of skewed data," ACM SIGKDD Explorations Newsletter, vol. 6, no. 1, pp. 50–59, 2004.

Chapter 8
Analytics Assisted Self-Optimization in LTE Networks

This chapter concentrates on self-optimization in wireless communication network and primarily introduces SON (Self-Organizing Networks) technology. Self-Organizing network (SON) is an automation technology which is designed to make the planning, deployment, operation, optimization and healing of mobile radio access networks simpler and faster. As the complexity of networks increases and an insatiable demand for mobile broadband continues, the need for SON has never been greater. It is considered as a major necessity in LTE and future 5G networks and operations due to the possible savings in capital expenditure (CAPEX) and operational expenditure (OPEX) by introducing SON in all phases of the network engineering life cycle.

This chapter introduces an application characteristic driven system, APP-SON, which is designed to achieve self-optimization in LTE networks through a wireless analytics assisted algorithm portfolio, and also discusses how APP-SON works in LTE network. APP-SON achieves targeted optimization through Profiling Cell Application Characteristics. Hungarian Algorithm Assisted Clustering (HAAC) algorithm and deep learning assisted regression algorithm are developed to cluster the 4G LTE cells via investigating cell level application characteristics.

8.1 SON (Self-Organizing Network)

As an automation technology in all phases of the network engineering life cycle: planning, deployment, and operation, SON is not just about the network, it is about the customer and quality of experience (QoE) they receive. It leaves operators free to focus on providing an excellent user experience since the management of technologies and vendor networks is simplified and streamlined. With good deployment, operators can get significant improvements in performance, better service quality, and increased revenues, all with fewer resources and minimum user touch.

© Springer International Publishing AG, part of Springer Nature 2018   107
Y. Ouyang et al., *Mining Over Air: Wireless Communication Networks Analytics*,
https://doi.org/10.1007/978-3-319-92312-3_8

Functionally, SON can also be divided into three major sub-functional groups as: Self-configuration functions in the deployment phase of the network engineering life cycle, Self-optimization functions in the phase of operation, and Self-healing functions in the phase of planning. SON is also categorized into centralized SON, Distributed SON and Hybrid SON by 3GPP [1, 3, 11–13]. In a centralized SON architecture, the algorithms are executed at the network management level. Commands, requests and parameter settings flow from the network management level to network elements, while measurement data and reports flow in the opposite direction. The SON algorithms run in the network nodes and the nodes exchange SON related messages directly with each other in the distributed SON architecture. In the hybrid SON architecture, a part of the SON algorithms run on the network management level and the remaining parts run in the network elements. It is an attempt to combine the advantages of centralized and distributed SON solutions, with the centralized coordination of SON functions and the ability to respond quickly to changes at the network element level. In the literature, there exist a rather vast collection of works on the SON. The IST project SOCRATES [2] has given several contributions to the understanding and development of SON. Load balancing and cell outage management are studied in [5, 6]. 4G Americas, next generation mobile networks (NGMN) and 3rd Generation Partnership Project (3GPP) have provided reports [7, 8, 10] where the most basic use cases of SON are discussed. Many future challenges are also discussed in works [4–9].

In SON, the trend in network operations is to gradually move from semi-manual toward autonomous planning, deployment, and optimization as Fig. 8.1 shows. A semi-manual operation means that SON functionalities suggest configurations which are first approved by the operator before being implemented. Autonomous network operation means that approval by the operator is skipped. In the planning phase, the

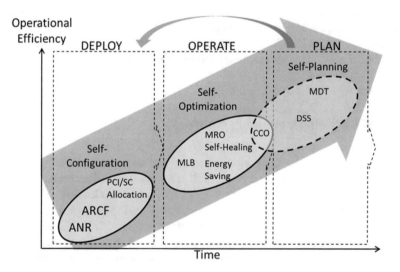

Fig. 8.1 Evolution of SON features

Centralized Coverage and Capacity Optimization (CCO) and decision support system (DSS) functions with the support of minimization of drive test (MDT) are implemented to reduce the operator's effort in planning. In the deployment phase, the self-configuration functions enable operators to install new nodes in plug-and-play fashion. Operators' effort in configuring and optimizing intra and inter LTE neighbors is reduced by the Automatic Neighbor Relations (ANR). Additionally, the effort in Physical Cell ID (PCI) allocation is also avoided. In the operating phase, the distributed SON functions, e.g. mobility robustness optimization (MRO), mobility load balancing (MLB), Energy Saving, enable operators to have cell-specific and dynamic configuration in contrast to base station based configurations in the planning phase.

8.2 APP-SON

3GPP [10–17] initiated the work towards standardizing self-optimizing and self-organizing capacities for LTE, and has already defined many use cases for Self-configuration, Self-optimization and Self-healing. SON vendors also claimed that their products could support all SON functions. However, Mobile operators need to deploy network neutral SON schemes and vendor neutral SON solutions. Mobile operators have specific and customized requirements on Self-Configuration, Self-Optimization and Self-Healing which are not fully supported in vendor specific SON solutions. 3GPP specifies use cases of handover and random access in self-optimization in SON, but has not yet any targeted solution to improve wireless performance and further user Quality of Experience. 3GPP also has not proposed any feasible proposal for self-planning and has not yet defined any practical algorithms to implement SON. Furthermore, more practical problems in 3GPP SON have not been addressed. For example, among thousands of performance indicators for each cell, how to prioritize those KPIs for optimization purpose? What are the traffic characteristics and patterns for each cell? Shall we profile the cell based on its traffic nature to find the corresponding and targeted performance indicators for us to optimize with high priority? Will the user Quality of Experience be improved after those KPIs are optimized? How to tune engineering parameters to optimize wireless KPIs and further to improve user Quality of Experience?

To address these challenges, an application characteristics driven system, APP-SON is designed to achieve self-optimization in LTE networks through a wireless analytics assisted algorithm portfolio. APP-SON achieves targeted optimization through Profiling Cell Application Characteristics. A Hungarian Algorithm Assisted Clustering (HAAC) algorithm and deep learning assisted regression algorithm are developed to cluster the 4G LTE cells via investigating cell level application characteristics. The homogenous application characteristics of cells in the same cluster are identified to prioritize target network KPIs in each cluster for best optimizing user Quality of Experience. Incremental learning scheme is also applied in temporal space in APP-SON. Cell based application nature is studied

Fig. 8.2 System
architecture of APP-SON

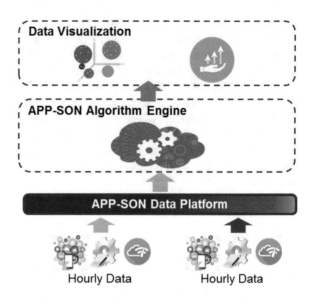

independently per hour since individual cells' application characteristics varies from
hours. App-SON then establish hourly cause effect between cell engineering parameters and target network KPIs through deep learning neural networks. For example,
video related performance indicators would be selected and optimized with a greater
priority for a cell where video traffic makes up more than 90% of the total traffic.
Finally a similarity based parameter tuning model in APP-SON is developed to
automatically tune cell engineering parameters per hour to optimize target network
KPIs and further to improve user Quality of Experience. The network performance
and user experience indicators could be instantly optimized by tuning the engineering parameters under the guide of the model created for a cell at a specific hour.
Through the APP-SON system, the targeted KPIs and user QoE can be optimized, or
improved close to target at an optimal level (Fig. 8.2).

8.3 APP-SON Architecture

The practical goal of APP-SON is to improve the targeted network KPIs and user
experience through tuning the corresponding engineering parameters in an incremental manner. The sample data is split into different groups by hour. In each group,
cells are further clustered into different clusters according to the patterns of application usage. Cells could be marked with clustering labels in the step of clustering.
However, the measured DPI (Deep Packet Inspection) data are collected from
multiple days in our experiment, which means that a cell might be marked with
different labels in the same hour from different days. Hence, those multiple labels
need to be combined to a single label for us to implement one corresponding
optimization policy in the same hour in the future.

There are five steps in the APP-SON system as shown in Figs. 8.3 and 8.4. The preliminary step is to split data into 24 groups according to the time when they are collected. For each group, all data (cells) are clustered into different clusters based on the patterns of application usage in the second step. The next step is to create regression model for each cluster, calculate the regression accuracy for each cluster, and determine the optimal number of clusters. The fourth step is to combine labeling results from different days. After building the APP-SON system, a labeling state diagram will be created for each cell in the LTE network. Such state diagram is able to show which cluster a particular cell should be located in at a specific hour in a day. In the final step, the engineering parameters will be automatically tuned to optimize the performance indicators and user quality of experience by using the similarity based parameter tuning system.

8.4 APP-SON Algorithm

In this section, we introduce how to utilize ensemble clustering algorithm to cluster and label cells, as well as deep neural network based regression method, and similarity based parameter tuning system in APP-SON. We will also go into greater depth on how to implement feedback based method to determine the best number of clusters and leverage combinatorial optimization algorithm to combine temporal clustering results (Table 8.1).

8.4.1 Hungarian Algorithm Assisted Clustering (HAAC)

Clustering is the task of grouping a set of objects in such a way that objects in the same cluster are more similar to each other than to those in other clusters. In the APP-SON, all cells should be clustered into different clusters according to the patterns of their application usage after the first step. In each cluster, cells are similar to each other in terms of the application usage. Clustering algorithms can be categorized based on their cluster model. Traditionally, those algorithms can be categorized into [18–21]: Connectivity-based clustering, Centroid-based clustering, Distribution-based clustering, and Density-based clustering. However, there is no objectively correct clustering algorithm. Every clustering algorithm implicitly or explicitly assumes a certain data model and it may produce erroneous or meaningless results when these assumptions are not satisfied by the sample data. The exploratory nature of clustering tasks demands efficient methods that would benefit from combining the strengths of many individual clustering algorithms. Therefore, it is meaningful to leverage ensemble clustering method which combines multiple clustering results generated by using different clustering methods or implementing a same clustering method many times with different initial parameters [22–29]. Instead of using basic clustering method, e.g. K-means, Fuzzy C-Means, GMM, or

---

INPUT:
Date records with A types of application and B performance
indicators for C cells in D days

OUTPUT:
L_{tc} diagram of labelling trend at clock t for c-th cell
$P_{j-1}, P_{j-2}, P_{j-3}, ..., P_{j-r}$: The value of tuning parameters

---

1: **Split training data set to 24 groups of hourly data**
2: **for** $t = 0{:}23$ **do**
3:    **for** $k = 3{:}20$ **do**
         // Clustering
4:         R_1 = Kmeans (k, application features)
5:         R_2 = FCM (k, application features)
6:         R_3 = GMM (k, application features)
7:         R_4 = HClust (k, application features)
8:         R_5 = GLARA (k, application features)
9:         R_k = Ensemble Clustering($R_1 R_2 R_3 R_4 R_5$)
         //DNN based regression
10:        $A_{all,k} = 0$
11:        **for** $h = 1{:}k$
12:            $y_h = f(X_1, X_2, ..., X_n)$
13:            $SS_{E,h} = (y, \hat{y})$
14:            $SS_{T,h} = (y, \hat{y})$
15:            $R_h = 1 - (SS_E / SS_T)$
16:            $A_{all,k} = (R_h)*W_h + A_{all,k}$
17:        **end for**
18:    **end for**
19:    $K_t = k$ (when $A_{all,k}$ = MAX ($A_{all,3}, ... , A_{all,20}$))
20:    **for** $c = 1{:}C$ **do**
21:        $L_{t,c}$ = Ensemble Clustering(R_{Kt})
22:    **end for**
    //Similarity based parameter tuning
23:    min_value = ∞
24:    t=1
25:    **for** $i = 1{:} K_t$ **do**
26:        **for** $j = 1$: Number of data points in i-th cluster **do**
27:            **IF** ($I_{Target} = I_j$)
28:                Dif = MIN (min_value, 1-Similarity$_j(P_j,P_c)$)
                //Calculated by using equation 2
29:                IF (Dif < min_value)
30:                    min_value = Dif
31:                    t = j
32:        **end for**
33:        $P_{j-1}, P_{j-2}, P_{j-3}, ..., P_{j-r}$
34:    **end for**
35: **end for**

Fig. 8.3 Algorithm of APP-SON

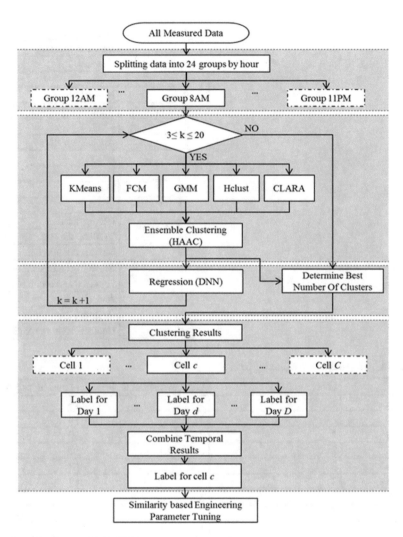

Fig. 8.4 Flow chart of APP-SON

Hierarchical clustering, we utilized ensemble clustering method for clustering analysis in the App-SON system. Specifically, Hungarian algorithm is employed as the ensemble method to relabel multiple clustering results generated by different clustering methods on the same data.

We first combine results from method A and B via relabeling labels in B by reference to A. To implement the Hungarian method in this case, the connection between these two results should be created. The cell 1 in Table 8.2 is marked with the same label "1" in both A and B. So the label "1" is very likely to be the same clustering label in both methods. In the Fig. 8.5a, we then connect 1 at the left side which represents the labels in method A to the 1 at the right side which represents the

Table 8.1 Features for clustering

Index	Applications
1	Instant messaging
2	Reading
3	Weibo
4	Navigation
5	Video
6	Music
7	Application store
8	Game
9	Online payment
10	Cartoon
11	Email
12	P2P application
13	VOIP application
14	Multimedia messaging
15	Browning and Downloading
16	Financial and economics
17	Anti-virus
18	Other applications
19	Unidentifiable applications

Table 8.2 Three different clustering results

	Label in method A	Label in method B	Label in method C
Cell 1	1	1	4
Cell 2	1	1	4
Cell 3	1	2	3
Cell 4	2	2	3
Cell 5	2	3	3
Cell 6	3	1	2
Cell 7	3	2	2
Cell 8	4	3	1
Cell 9	4	4	1

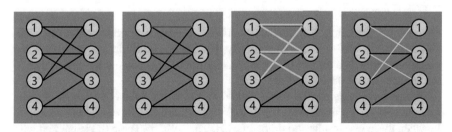

Fig. 8.5 The process of implementing Hungarian algorithm

labels in method B. Cell 2 is same as cell 1, we do the same thing to connect 1 and 1. Using the same way, cell 3 is labeled as 1 in method A and 2 in method B, so 1 at the left side should also be linked to 2 at right side. After implementing the same way to all nodes, the connection figure could be made as Fig. 8.5a which shows the relationship between labels generated by method A and B. Such links between the two sides are just potential links. In Fig. 8.5a, each node at the left side is connected to many nodes at the other side. After implementing the Hungarian algorithm, each node at the left side will be only connected to one node at the right side.

By implementing the Hungarian algorithm, 1 at the left side can be linked with 1 at right side, and 2 at left side can be linked to 2 at right side using red edge as Fig. 8.5b shows, since they have potential connections. Such red edges are considered as matched edges and remaining black edges are unmatched edges (they are still potential connections). But 3 at left side cannot be linked to either 1 or 2 at the right side, since both nodes at the right side are previously occupied by 1 and 2 from the left side (the red edges). In this case, an augmenting path should be made from node 3 at the left side. In Hungarian algorithm, an augmenting path is defined as a path in which the edges start with an unmatched edge and belong alternatively to matched edges and unmatched edges. In this case, from node 3 at the left side, an augmenting path could be created as L3- > R1- > L1- > R2- > L2- > R3, L3 means the node 3 at left side, R1 means the node 1 at right side. After creating the augmenting path, the bright green edges in Fig. 8.5c, we try to find edges which are matched edge and also on the augmenting path. In Fig. 8.5c, it is clear that edges L1-R1 and L2-R2 are such edge with two colors. According to the Hungarian algorithm, we remove such edges with two colors and finally get the links between the left side and right side. In Fig. 8.5d, L1 is linked with R2, L2 is linked with R3, and L3 is linked with R1. For L4, it can be linked with R3, but R3 is occupied by L2, so it keep trying other potential links and find R4 is available to be linked. Figure 8.5d is the result of Hungarian algorithm, it shows that label 1 at right side should be considered as label 3 at left side, label 2 should be considered as label 1 at left side, label 3 at right side should be considered as label 2 at left side, and label 4 at right side can be considered as 4 at left side. Table 8.3 shows the result of relabeling using Hungarian algorithm, 1 in method B is relabeled as 3, 2 in method B is relabeled as 1, 3 in method B is

Table 8.3 Relabeling and voting

	Label in method A	Label in method B	Label in method C	Final (Voted) label
Cell 1	1	3	1	1
Cell 2	1	3	1	1
Cell 3	1	1	2	1
Cell 4	2	1	2	2
Cell 5	2	2	2	2
Cell 6	3	3	3	3
Cell 7	3	1	3	3
Cell 8	4	2	4	4
Cell 9	4	4	4	4

relabeled as 2, and 4 in method B is still 4 without any change. Using the Hungarian algorithm, clustering results from method B and C could be relabeled based on a reference result which is A in this case. After voting among three relabeled result, the final labeling result should be the right column in Table 8.3, and each cell in the data set is only marked with one label.

8.4.2   Units Regression Assisted Clustering Number Determination

Since our goal is to create regression model for each cluster, the accuracy of regression could be impacted by the clustering result. It means that a good clustering algorithm could suitably cluster data into different clusters on which the regression model could have better accuracy. In this case, to determine the optimal number of clusters, it is very meaningful to use the regression accuracy as the evaluation criterion to determine the number of clusters. In APP-SON system, we set the number of clusters from 3 to 20, and calculate the regression accuracy for each cluster very time. The number which makes the best regression accuracy will be considered as the optimal number of clusters. The regression accuracy, Aall, is the weighted sum of regression accuracies in all clusters and calculated by Eq. 8.1 as follow:

$$A_{all} = \max\left\{\sum_{i=1}^{k} A_{k,i} W_i, k \in [3, N]\right\} \tag{8.1}$$

Where $A_{k,i}$ is the regression accuracy for i-th cluster using K as the number of clusters. W_i is the weight and calculated as the percentage of samples in the i-th cluster. N is the maximum number of clusters tried in this method and set as 20 in the experiment. For example, 5 would be the optimal number of clusters if $A_{all} = \sum_{i=1}^{5} A_{5i}W_i$, which means that the weighted sum of regression accuracies using 5 as the number of clusters is the maximum one among all cases using the number of clusters from 3 to 20. So 5 is the optimal number of clusters.

8.4.3   DNN Based Regression

Regression is a statistical process for estimating the relationships among variables. It includes techniques for modeling and analyzing several variables, when the focus is on the relationship between a dependent variable and one or more independent variables or predictors. Traditionally, regression methods can be categorized into linear and non-linear regression [30–34]. Linear regression concerns two-dimensional sample points with one independent variable and one dependent variable. It finds a linear function and predicts the dependent variable values as a function of the independent variables. On the contrary, nonlinear regression is a form of regression analysis in which observational data are modeled by a function which is a nonlinear combination of the model parameters and depends on one or more independent variables. The data are fitted by a method of successive approximations. In APP-SON, we utilize deep neural network (DNN) as regression method to derive the relationship between engineering parameters and performance indicators, since DNN contains multiple layers of neural networks which use activation function to implement the functionality of nonlinear regression and derive the relationship between input variables and output variables more accurately.

As Fig. 8.6 shows, x stands for input variables, it can be features passed forward from the DNN's previous layer, and will be fed into each node of the next hidden layer. Each x will be multiplied by a corresponding weight w. The sum of those products is added to a bias, b, and fed into an activation function. The activation function, rectified linear unit (ReLU), is used in this system since it doesn't saturate on shallow gradients as sigmoid activation functions do. The layer of neural network

Fig. 8.6 Regression using deep neural network

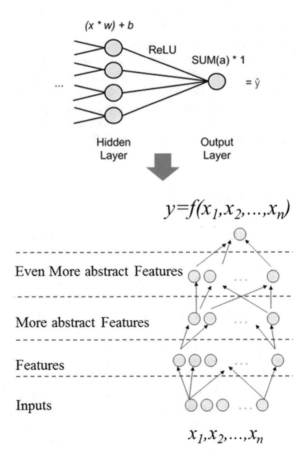

performing regression has one output node, and that node will multiply the sum of the previous layer's activations by 1. The result is $\hat{y}$ which is the dependent variable that all the input x map to. To perform backpropagation and make the network learn, the ground-truth value of y is compared with the predicted value $\hat{y}$. The weights and biases of the network are adjusted until the difference between y and $\hat{y}$ is minimized. A deep neural network has multiple hidden layers of units between the input and output layers, which is able to model complex non-linear relationships. The extra layers enable composition of features from lower layers as the bottom part shows in Fig. 8.6. The output of the previous layers is considered as more abstracted features and taken as the input of the next layer.

8.4.4   Combining Labels in Temporal Space for Each Cell

Combinatorial optimization is an approach to find the best solution out of a very large set of possible solutions. Traditionally, combinatorial optimization algorithm can be leveraged to develop the best airline network of spokes and destinations, decide which taxis in a fleet to route to pick up fares, determine the optimal way to deliver packages, or determine the right attributes of concept elements prior to concept testing. In the ensemble clustering [23, 24], it is popularly used to combine the clustering results generated from different clustering methods or a same method with different parameters. In APP-SON, a cell at each hour should only be marked with one label. Inspired by the combinatorial optimization based ensemble clustering, we utilize the combinational optimization algorithm in APP-SON to reduce the dimensions of temporal data from different days.

8.4.5   Similarity Based Parameter Tuning

After clustering, a cell can be clustered into a cluster and marked with a label. Since the data is collected from multiple days, a cell might be marked many times with different labels. It is hard to determine which label a cell should be marked with. Figure 8.7 shows that cells from 1 to N are marked with different labels at each hour every day due to the different application usages in each hour. The experimental data is collected from M different days, and causes M different labeling results like the color plates in Fig. 8.7. Each column on these color plates can be combined by implementing the Hungarian algorithm. The final result would be the right one in Fig. 8.7. It is the final labeling result for each cell at each hour. After labeling each cell, the relationships between engineering parameter and performance indicators can be derived for each cluster. As the algorithm in Fig. 8.3 shows, there are r engineering parameters and N data records in a cluster. The targeting value for a performance indicator is given as input for this algorithm. Among all data points which have the same value of performance indicator with the targeting value are

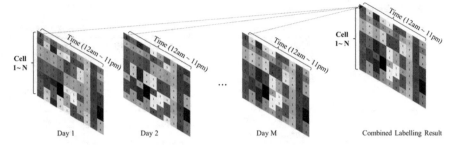

Fig. 8.7 Combine temporal results from different days

selected as candidate data points. Among these candidate data points, the data point which has the highest similarity with the one waiting to be tuned is considered as the reference to tune the engineering parameters. The similarity is calculated by using Eq. (8.2):

$$Similarity_i = 1 - \sum\nolimits_{n=1}^{r} \frac{|P_{i-n} - P_{c-n}|}{R_n} \tag{8.2}$$

where P_{i-n} is the value of n-th parameter for the i-th data record in this cluster. P_{c-n} is the value of n-th parameters for the current cell. R_n is the value range of the n-th parameter.

8.5   Simulation and Discussion

The sample data is collected from 840 cells of a tier one wireless operator from Nov. 1st to Dec. 31st 2016, including urban and rural area. The data set includes 60,480 hourly data records, and is composed of contents listed in Table 8.4. The 60,480 data records are first split into 24 groups according to the time of collection. There are 2530 data records in each group which include 3 days' information of 840 cells at that hour. The R squared value is utilized as the regression accuracy of DNN based regression in the simulation. Figure 8.8 shows the results calculated by using Eq. 8.1 on four groups of data set, 12 AM, 8 AM, 12 PM and 8 PM. The ensemble clustering is implemented on these four groups of data with the number of clusters from 3 to 20. It is clear that the optimal number of clusters is 5 for the group of data collected at 12 AM, 9 for the groups of data collected at 8 AM, 20 for data collected at 12 PM and 18 for data collected at 8 PM.

After determining the optimal number of clusters, the data in each group will be clustered according to the patterns of application usage. A cell at each group will be marked with a clustering label after the clustering analysis showing the feature of this cluster. Figure 8.9 shows the labeling results for the group of data collected at 12 AM. Through the previous section, it is known that the optimal number of

Table 8.4 Three different clustering results

Type	Data content	Type	Data content
Cell attribute	Cell ID	Application traffic (19)	Instant Messaging
	Region		Reading
	Type of location (urban/ rural)		Social Media
	Number of users		Navigation
	Number of devices		Video
Date/time	Date		Music
	Time (12AM ~ 11PM)		Application store
Engineering parameters	Azimuth angle		Game
	Pitch angle		Online payment
	Height		Anime
	Longitude		Email
	Latitude		P2P application
	Reference signal power		VOIP application
	M-DownTilt		Multi-media messaging
	E-DownTilt		Browning and downloading
	Others (30 parameters)		Finance
Network performance indicator (8 application related KPI)	TCP successful connection rate		An ti-virus
	HTTP successful connection rate		Other applications
	Video connection request successful rate		Unidentifiable applications
	Application downloading service request successful rate	Network performance indicator (4 General KPI)	Network connection Rate
	Instant messaging service request successful rate		Dropped call Rate
	TCP establishment latency		RRC connection Rate
	HTTP session latency		ERAB connection Rate
	Web page display latency		

clusters for this group is 5. According to the proportion of application traffics, these five clusters can be marked with labels as "Video/ Others/ Browsing & Downloading", "Video/ Browsing & Downloading/ Others/Reading", "Video/

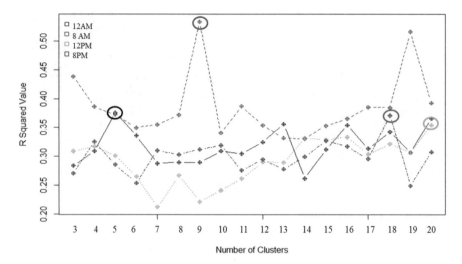

Fig. 8.8 Optimal number of clusters

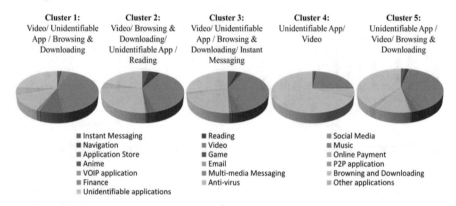

Fig. 8.9 Clustering labels for group of 12 AM

Others/ Browsing & Downloading/Instant Messaging", "Others/ Video", and "Others/Video/Browsing & Downloading". After combining different labeling results from multiple days, each cell is only marked with one single label for each hour. Since the patterns of traffic usage vary from hour to hour, the process of clustering and regression analysis is independent from each hour. Figure 8.10 shows the changes of clustering label of a cell from 12 AM to 11 PM in a day, which indicate the application behavior change for this cell at 1 day. That requests us to further find targeted KPIs in each hour to optimize. For each cluster in a group, DNN is used to determine the number of clusters and can also be used to derive the relationship between engineering parameters and performance indicators. However, the R squared values of the regression in all groups are very low, which means that the vast majority of data points are distributed far from the fitted regression line.

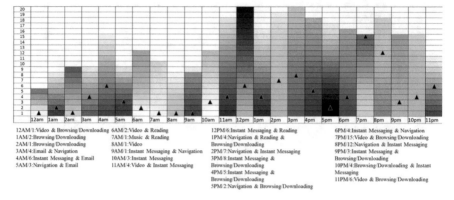

12AM/1:Video & Browsing/Downloading	6AM/2:Video & Reading	12PM/6:Instant Messaging & Reading	6PM/4:Instant Messaging & Navigation
1AM/2:Browsing/Downloading	7AM/1:Music & Reading	1PM/4:Navigation & Reading &	7PM/15:Video & Browsing/Downloading
2AM/1:Browsing/Downloading	8AM/1:Video	Browsing/Downloading	8PM/12:Navigation & Instant Messaging
3AM/4:Email & Navigation	9AM/1:Instant Messaging & Navigation	2PM/7:Navigation & Instant Messaging	9PM/3:Instant Messaging &
4AM/6:Instant Messaging & Email	10AM/3:Instant Messaging	3PM/8:Instant Messaging &	Browsing/Downloading
5AM/3:Navigation & Email	11AM/4:Video & Instant Messaging	Browsing/Downloading	10PM/4:Browsing/Downloading & Instant
		4PM/5:Instant Messaging &	Messaging
		Browsing/Downloading	11PM/6:Video & Browsing/Downloading
		5PM/2:Navigation & Browsing/Downloading	

Fig. 8.10   Hourly labels per cell

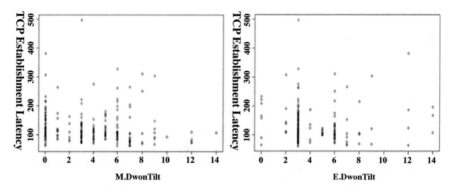

Fig. 8.11   Scatter plot for engineering parameters and performance indicator (TCP Delay) at 8 AM

Instead of drawing regression line, we draw the scatter plot for each pair of engineering parameter and performance indicator and analyze the distribution of data points.

Two groups of data, 12 AM and 8 AM, which represent two different hours with low and high data traffic, are studied respectively in the experiment. The scatter plots in Fig. 8.11 show that data points are irregularly scattered, which means that there is no parameter to tune. On the contrary, the data points with high TCP Establishment latency are only scattered at a specific range of engineering parameters in Fig. 8.12. For example, data points with TCP Establishment latency value higher than 1000 are only present at the range between 3 and 7 in M.DownTilt and 3 to 6 in E.DownTilt. It means that the TCP Establishment latency can be decreased by setting the M. DownTile and E.DownTile. The results of the experiment are meaningful, since the data traffic is very low at 12 AM. Even though the cells are clustered into different clusters, the patterns are not much separable with each other. The relationship between engineering parameters and performance indicators for the group of 12 AM is not correctly derived. This situation is changed at the group of 8 AM. Data

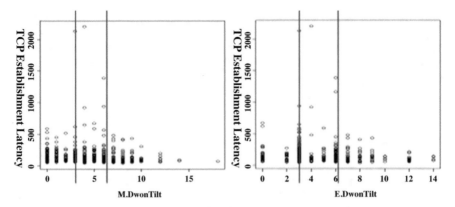

Fig. 8.12 Scatter plot for engineering parameters and performance indicator (TCP Delay) at 12 AM

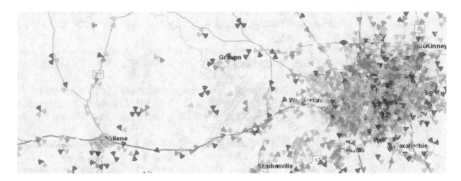

Fig. 8.13 Cells labelling

traffic is increased and relationships between engineering parameters and performance indicators become recognizable as Fig. 8.12 shows. Figure 8.13 shows the labelling result of cells within an area on the APP-SON system. Different colors represent different clusters. For each cluster, a corresponding parameter tuning policy is defined and enforced to automate self-optimization. The network performance and user quality of experience in the incoming hours can be timely tuned under the guidance of these policies.

APP-SON leverages scalable big data platform for targeted optimization through profiling cell application characteristics in an incremental manner in temporal space. A Hungarian Algorithm Assisted Clustering (HAAC) algorithm and deep learning assisted regression algorithm are developed to profile the cell application characteristics and find the targeted KPIs to be optimized for each cell. A similarity based parameter tuning algorithm is developed to tune the corresponding engineering parameters to optimize the targeted KPIs which further improve QoE. Experiment results demonstrate that the APP-SON system can precisely profile cell traffic and application characteristics to find the targeted KPIs for optimization for each cell.

APP-SON can also automatically tune the corresponding engineering parameters to improve the corresponding KPIs to ultimately improve QoE. APP-SON has been successfully implemented in production and applied in a tier-1 operator's 4G network and as universal SON solution it will be smoothly transitioned and applied in 5G networks for this operator.

References

1. Hämäläinen, Seppo, Henning Sanneck, and Cinzia Sartori. LTE self-organising networks (SON): network management automation for operational efficiency. John Wiley & Sons, 2012.
2. EU FP7 project SOCRATES homepage, http://www.fp7-socrates.org.
3. M. Amirijoo, R. Litjens, K. Spaey et al., "Use cases, requirements and assessment criteria for future self-organising radio access networks," in Proceedings of the 3rd International Workshop on Self-Organizing Systems(IWSOS'08), pp. 10–12, Vienna, Austria, December 2008.
4. L. C. Schmelz, J. L. van den Berg, R. Litjens et al., "Self-organisation in wireless networks—use cases and their interrelations," in Proceedings of the 22nd Wireless World Research Forum Meeting, Paris, France, May 2009.
5. A. Lobinger, S. Stefanski, T. Jansen, and I. Balan, "Load balancing in downlink LTE self-optimizing networks," in Joint Workshop COST 2100 SWG 3.1 & FP7-ICT-SOCRATES, Athens, Greece, February 2010.
6. M. Amirijoo, L. Jorguseski, T. Kürner et al., "Cell outage management in LTE networks," in Proceedings of the 6th International Symposium on Wireless Communication Systems (ISWCS'09), pp. 600–604, Siena, Italy, September 2009. View at Publisher · View at Google Scholar · View at Scopus
7. NGMN, "Use Cases related to Self Organising Network," Overall Description Release Date: May, 2007 http://www.ngmn.org/nc/downloads/techdownloads.html.
8. 4G Americas, "Self-Optimizing Networks—The Benefits of SON in LTE," http://www.4gamericas.org/, 2011.
9. N. Marchetti, N. R. Prasad, J. Johansson, and T. Cai, "Self-organizing networks: state-of-the-art, challenges and perspectives," in Proceedings of the 8th International Conference on Communications (COMM'10), pp. 503–508, June 2010. View at Publisher · View at Google Scholar · View at Scopus
10. Feng, Sujuan, and Eiko Seidel. "Self-organizing networks (SON) in 3GPP long term evolution." Nomor Research GmbH, Munich, Germany (2008): 1–15.
11. Peng, Mugen, et al. "Self-configuration and self-optimization in LTE-advanced heterogeneous networks." IEEE Communications Magazine 51.5 (2013): 36–45.
12. Ramiro, Juan, and Khalid Hamied, eds. Self-organizing networks (SON): self-planning, self-optimization and self-healing for GSM, UMTS and LTE. John Wiley & Sons, 2011.
13. L. Jorguseski, A. Pais, F. Gunnarsson, A. Centonza and C. Willcock, "Self-organizing networks in 3GPP: standardization and future trends," in IEEE Communications Magazine, vol. 52, no. 12, pp. 28–34, December 2014.
14. Ouyang, Ye, et al. "A novel methodology of data analytics and modeling to evaluate LTE network performance." 2014 Wireless Telecommunications Symposium. IEEE, 2014.
15. Clímaco, João, and José Craveirinha. "Multicriteria analysis in telecommunication network planning and design—problems and issues." Multiple Criteria Decision Analysis: State of the Art Surveys. Springer New York, 2005. 899–941.
16. Ouyang, Ye, and Tan Yan. "Profiling Wireless Resource Usage for Mobile Apps via Crowdsourcing-Based Network Analytics." IEEE Internet of Things Journal 2.5 (2015): 391–398.

17. Deb, Supratim, and Pantelis Monogioudis. "Learning-based uplink interference management in 4G LTE cellular systems." IEEE/ACM Transactions on Networking (TON) 23.2 (2015): 398–411.
18. Carbonell, Jaime G., Ryszard S. Michalski, and Tom M. Mitchell. "An overview of machine learning." Machine learning. Springer Berlin Heidelberg, 1983. 3–23.
19. Alnwaimi, Ghassan, et al. "Machine learning based knowledge acquisition on spectrum usage for lte femtocells." Vehicular Technology Conference (VTC Fall), 2013 I.E. 78th. IEEE, 2013.
20. Rousseeuw, Peter J. "Silhouettes: a graphical aid to the interpretation and validation of cluster analysis." Journal of computational and applied mathematics 20 (1987): 53–65.
21. Tibshirani, Robert, Guenther Walther, and Trevor Hastie. "Estimating the number of clusters in a data set via the gap statistic." Journal of the Royal Statistical Society: Series B (Statistical Methodology) 63.2 (2001): 411–423.
22. Strehl, Alexander, and Joydeep Ghosh. "Cluster ensembles--knowledge reuse framework for combining multiple partitions." Journal of machine learning research 3.Dec (2002): 583–617.
23. Topchy, Alexander, Anil K. Jain, and William Punch. "Clustering ensembles: Models of consensus and weak partitions." IEEE Transactions on pattern analysis and machine intelligence 27.12 (2005): 1866–1881.
24. Meilă, Marina. "Comparing clustering: an axiomatic view." Proceedings of the 22nd international conference on Machine learning. ACM, 2005.
25. Amigó, Enrique, et al. "A comparison of extrinsic clustering evaluation metrics based on formal constraints." Information retrieval 12.4 (2009): 461–486.
26. Jain, Anil K. "Data clustering: 50 years beyond K-means." Pattern recognition letters 31.8 (2010): 651–666.
27. Bezdek, James C., Robert Ehrlich, and William Full. "FCM: The fuzzy c-means clustering algorithm." Computers & Geosciences 10.2–3 (1984): 191–203.
28. Johnson, Stephen C. "Hierarchical clustering schemes." Psychometrika 32.3 (1967): 241–254.
29. Ester, Martin, et al. "A density-based algorithm for discovering clusters in large spatial databases with noise." Kdd. Vol. 96. No. 34. 1996.
30. Asai, H. Tanaka-S. Uegima-K. "Linear regression analysis with fuzzy model." IEEE Trans. Systems Man Cybern 12 (1982): 903–907.
31. Duggleby, Ronald G. "Regression analysis of nonlinear Arrhenius plots: an empirical model and a computer program." Computers in biology and medicine 14.4 (1984): 447–455.
32. BRKSPM-2005 – Cisco Self-Optimizing Network (SON) Architecture (2014 Milan)
33. Ericsson automates optimization of mobile networks, PRESS RELEASE February 28, 2012
34. The Logical Rise of C-SON: Discover compelling reasons why vendor agnostic C-SON is critical in increasingly complex HetNets

Chapter 9
Introduction – Telecommunications Data and Marketing

Starting with this Chapter 9, the rest of this book is devoted to applications and studies specific to telecom marketing. Telecommunication data are of paramount importance; when combined with statistical and econometric models, and machine learning tools, it gives specific insights into consumer behavior. And with better knowledge of consumer behavior, companies (including mobile carriers) can carry out more effective marketing campaigns to better target consumers and enhance its sales performance. In this section of telecom marketing, we focus on four main dimensions: customer identification, customer attraction, customer retention, and customer development. The ideal result is to determine how to predict customer churn rate, to recommend best rate plans, and to develop new services to subscribers. The churn rate problem specifically accommodates customer attraction and retention.

In addition to data on individual profile and historic behavior, two additional features of telecommunication data make it valuable to consumer behavior research: one is the individual social network information inferred from the call detailed records that are automatically recorded by telecommunications companies, and the other is the individual location information recorded by the signaling data. These two types of information greatly extend the horizon of consumer behavior research and enable the possibility of studying the behavior taking into consideration of social network, i.e. individual interactions and geographic location, i.e. individual mobility.

With the wide adoption of telecommunication products (especially mobile phones) a tremendous amount of telecommunication data (including detailed information about individuals' characteristics, behavior and their social networks) allow us to explore many new questions regarding consumer behavior associated with network and social process. People use mobile phones to connect to the world and, more importantly, to other people. This presents wireless carriers and researchers with a great opportunity to study consumer behavior, social networks, and ultimately the interplay between the two. Previous studies in marketing have used the information about consumer characteristics and their interactions with the carriers, such

© Springer International Publishing AG, part of Springer Nature 2018
Y. Ouyang et al., *Mining Over Air: Wireless Communication Networks Analytics*,
https://doi.org/10.1007/978-3-319-92312-3_9

as their choice of plans and service use (e.g. [1–4]). In fact, mobile network data contain extensive longitudinal information at the individual level about the contact information which can be used to infer social network. Eagle et al. [5] compared the mobile phone data and the self-reported survey data; they found that these two sets of data can be distinct from each other, but that "the data collected from mobile phones can be used to accurately predict a cognitive construct such as friendship."

The network based on mobile data is constructed based on the call detail record, which is produced by telecommunications equipment automatically. The record includes the details of a phone call or other communications transaction (e.g., text message) that passes through that facility or device. The record contains various attributes of the call, such as time, duration, completion status, source number, and destination number, etc. There are a few literature starts to take advantage of the social network information based one mobile data. Onnela et al. [6] studied the local and global network structure of the communication network for mobile phone users and observed the correlation between tie strength and local network structure around the tie. Eagle et al. [7] combines the telecommunication data with the national census data and found the correlation between the diversity of network with social economic development. Godinho de Matos et al. [8] used IV approach to investigate social influence on iPhone 3G adoption using mobile data from Europe. The following chapters also provide three studies demonstrating how we use the social network data inferred by mobile contacts in consumer behavior research and what managerial problems can be solved using the information. They are based on three papers of Hu et al. [9] and Hu et al. [10, 11].

In addition to social network studies, the research on individual mobility is another much newer frontier of behavior research that only became available after the tremendous amount of mobile data became available. Gonzalez et al. [12] is one of the earliest works in this area. The authors studied the trajectory of 100,000 anonymized mobile phone users whose position is tracked for 6 months. They showed that individual trajectory is non-random and shows a high-degree of spatial-temporal regularity. Song et al. [13] demonstrated that a small fraction of the time-resolved location data is enough to recover the real mobility pattern. Palchykov et al. [14] built a model based on the aggregated mobile phone call data and made a good prediction on the human flow pattern. Compared to social network studies, the individual mobility study is still at an early stage. In marketing, no research can be found in this area so far, so it calls for more talent to work on this topic. Companies need the information to solve questions such as location selection and geographic based targeting.

New data engenders new research possibilities. However, analyzing such data is challenging. From a statistical perspective, consumer behavior models are derived from marketing and social science domain knowledge. They focus on individual-firm relationship and individual-individual interactions, taking advantage of data collected. Deciding on which models and strategies would largely depend on the nature of data retrieved. In this part, our prior interest will be on the structures and dynamics that are related to individual environment, their characteristics and social relationships. Through the studies, this part will summarize concepts of social interactions and network influences from graph structures. They focus on how those structures can be leveraged for marketing purposes and concrete business problems.

The following chapters focus on the research of individual behavior in the specific context of social networks. In Chap. 10, switching behavior of customers from one carrier to another one is analyzed to predict customer churn rate. The dynamic model integrates both social-learning and network effects into one single model, which allows us to compare effects of those two components. In Chap. 11, consumer adoption behavior for a new mobile phone is analyzed. Effects of network structures on individual adoption are incorporated and measured, looking for both local and global network effects. In Chap. 12, dynamic network structures and measures are used to predict social influence on the diffusion process of a new product.

9.1   Topics in Telecom Marketing

In customer relationship management, there are four main focuses: customer identification, customer attraction, customer retention, and customer development. With the help of telecom marketing, organizations can better discriminate and more effectively direct resources to the most profitable group of customers. Customer identification involves targeting the population who are most likely to become customers (or most profitable). After potential customers are identified, customer attraction focuses on targeting certain customer segments. An important part of customer relationship management is customer retention, particularly in saturated markets, which aims to retain customers by providing products and services that satisfy their expectations. Lastly, customer development takes this a step further to increase the value of those customers.

When analyzing customer behavior in telecom marketing, an important piece of information that can be utilized is individuals' social network. Incorporating such information into a proper model enables us to examine deeper hypothesis and increase credibility of results. Contrary to relational databases used in previous chapters, algorithms should take into account graph-network structure effects. In the first study presented in Chap. 8, those effects are modeled in a dynamic way to study churning (switching behavior from a mobile carrier to other ones), combining in particular spread behavior within network and social-learning from peers. Strategies for network exploration are also discussed in this chapter, by describing snowball-sampling and clustering identification of networks from a large connected graph.

In general, there are two mechanisms that explain why customers are influenced by other peers in their social network: the customer is either backward looking or forward looking. First, backward looking individuals use feedback from their peers to update their own expectations of product quality [15]. This is known as social learning. There are two types of social learning: word-of-mouth communication

[16], in which peers exchange information directly, and by observational learning [17, 18], in which observation of others' behaviour leads to an individual to infer that their peers may have some prior information that they do not possess. Word-of-mouth communication [19, 20] and observational learning [17, 18, 21, 22] are regarded as the two mechanisms that drive such a social-learning process.

The second mechanism that motivates customers to change their behaviour is direct benefit gained from the network when forward looking. In this case, people may want to follow others if direct utility can be gained by aligning their behavior with that of others; this is called the network effect. The network effect refers to particular forms of externalities such that the actions of peers (i.e., the network size) directly affect an individual's preference [23]. The network effect differs from the social-learning or informational effect in that others' behavior affects an individual's payoff directly, rather than indirectly by updating his or her information set.

Few papers have studied these two mechanisms together. Goolsbee and Klenow [24] were among the first to claim that network externalities and learning from others can be treated as local spillovers and that they affect the diffusion of home computers so that people are more likely to purchase their first home computer if a large share of their peers own computers. Using aggregate movie box-office data, Moretti [15] estimated the importance of the social-learning effect on consumer choices when the product quality is difficult to observe in advance. In his article, social learning will occur as result of direct communication with peers and/or observation of the peers' purchases.

Social influence can also affect diffusion of new products. This is the focus of both Chaps. 11 and 12. In particular, Chap. 11 proposes a new targeting strategy, network-based targeting and Chap. 12 presents the study of social influence.

Since individuals influence each other's behavior through social contact, the network structure that summarizes the nature of pattern of social interaction also influences user behavior. So we can use that as additional information to predict individuals' future behavior. The study on this topic is presented in Chap. 11. Figure 9.1 shows the adoption patterns in two networks extracted using telecom data. The black dots indicate individuals and the lines represent the connections between each dots. They are of the same size (about 200 individuals) in the same city, the only difference being their network structure. Using the same network graphing algorithm, one network looks like a tight string ball while the other has a "hub-and-spoke" appearance. Over the same period, 24 individuals (denoted as red dots) adopted the new product in the tight pattern whereas there were only 10 in the "hub-and-spoke" pattern. One factor that explains the observed difference in behavior is the network structure. Network structure may affect individual adoption behavior via three separate channels. First, through own and contextual effects as a result of individual characteristics and positions within the network; secondly, through global network structures such as the environment (correlated effect); and thirdly, by peer influence (endogenous effect). Mobile data contain extensive information at the individual level, not just about behavior but also about social networks, and a dataset that combines both types of information provides opportunities for

Fig. 9.1 Network structure
and Samsung Note II
adoption

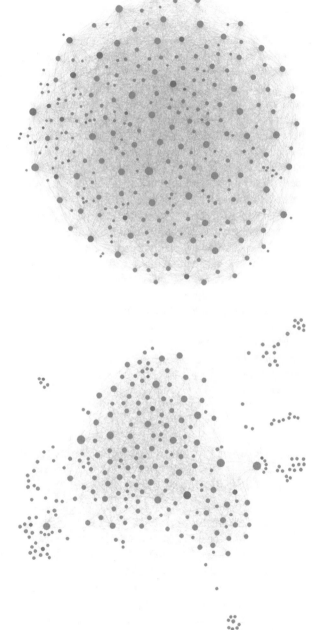

developing new models and generating new tools, which may lead to changes in the common industry practices. So as a result, we propose a new targeting strategy, NETWORK BASED TARGETING.

In Chap. 12, we narrow down our focus to a specific effect that can only be observed in networks, i.e. social influence (or peer effect). Social influence has been extensively studied due to the multiplier effect generated by social influence, which facilitates the product diffusion process. Social influence expresses an individual's conformity motive in adopting a behavior similar to that of his/her peers, particularly when that behavior is exhibited by many of those peers [25]. A network may or may not exhibit conformity, may have positive or "negative" conformity, depending on the individual interactions in that network. Well-connected individuals theoretically have a high propensity to be influenced by friends. Dense networks consequently have high chances of exerting social influence, which in turn results in strong behavior diffusion among network members.

In general, there is little literature on mobile data to take advantage of the social network information. In addition to the studies used in the book, Onnela et al. [6] studied the local and global network structure of the communication network for mobile phone users and observed the correlation between tie strength and local network structure around the tie. Eagle et al. [7] combined the telecommunication data with the national census data and found the correlation between the diversity of network and the social economic development. Godinho de Matos et al. [8] used IV approach to investigate social influence on iPhone 3G diffusion using mobile data from Europe. In summary, this social network information allows researchers to study consumer behavior and choices in the context of networks. It is a promising area that has great potential to advance people's knowledge in marketing and business.

9.2   General Construction of a Social Network

9.2.1   Collection and Types of Data

An initial phase before applying a model is to select, collect, clean and manage data. In general, the access to data is not unlimited and can be costly to produce or obtain. Some constraints and limitations are inevitable. For example, data are usually limited in space, in time and in extent. Consequently, one may choose what type of data is important to collect. Also, data may be collected from different sources that need to be merged conveniently. Finally, preparation of data includes extracting relevant parts of the data which will be effectively analyzed.

Generally, the whole data of interest cannot be obtained, and collected data are limited. For example, in telecom marketing problems, data are usually available for only one telecom carrier or one mobile phone brand. Secondly, some features of interest are inaccessible because they are related to individual environments, one's own personality or unknown social relationships. Furthermore, only a part of the structure and dynamics are obtained: collection is limited in space and time, and knowledge of social relationship between customers may be missing.

A lack of structured information can lead to several issues: when some factors specific to a group are lacking, high bias estimations may occur such as the correlated effect problem. This is when a complex model is designed with too few informative data: strong simplifications and assumptions may lead to statistical identification problems, such as the reflection problem (for example with linear-in-means social interaction model). Strategies to solve those problems using network local and global structures are explained in Chap. 11, focusing on the study of consumer adoption behaviors.

9.2.2   Extraction and Management of the Network

Eagle et al. [5] compared mobile network data and self-reported survey data and found that "the data collected from a mobile network can be used to accurately predict cognitive constructs such as friendship." This finding suggests that mobile network information would be useful in studying other consumers' influence on user behavior with respect to subscription to a wireless carrier.

There are three strategies for collecting measures of a network structure. The full network strategy takes into consideration the entire network. The ego network strategy begins with focal nodes and then tracks down the nodes that are directly connected to the focal ego. The snowball network strategy begins with a set of focal nodes. The contacts of the focal nodes and the contacts of the contacts are tracked down, and the process continues until the user decides to stop. The snowball network is an extended ego network. There is no "right" strategy for all problems. Snowball-sampling is often used to construct a social network from mobile data. Chen et al. [26] demonstrated that the snowball sampling method performs better in recovering social interactions because it better preserves the network structure. The sample was constructed as follows: seeds were randomly selected from the entire mobile network. Based on these seeds, we identified everyone they contacted by either call or text message over a short period of time at the beginning of the sample period. We considered those direct contacts to be the neighbors or "friends" of the seeds. Together, they formed a sphere of a network group. We then collected information about the direct contacts of the neighbors' neighbors. We thus formed a two-layer snowball for each seed. Complete network and demographic information for each individual is included in the sample. We repeat the process until we reach some predetermined target.

Diagrams help to better understand this sampling procedure. In Fig. 9.2a, we demonstrate how we specify the network in this study by showing the network topology in each step. We first randomly pick an individual i. There are five immediate nodes that are connected to i, which is now highlighted with a frame. We call those nodes "neighbors" for individual i. In Fig. 9.2b, we highlight all group

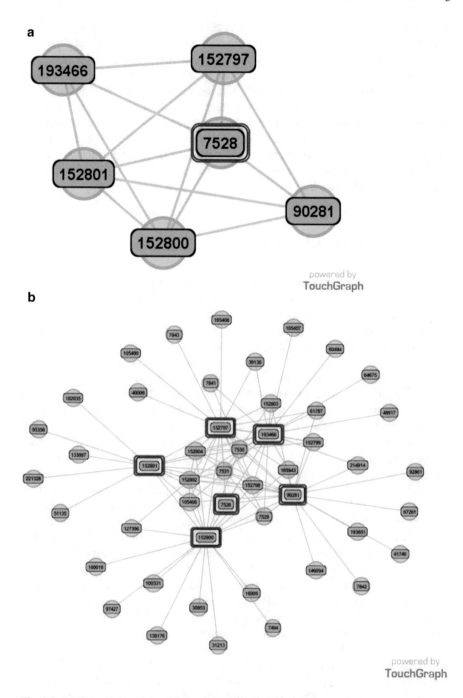

Fig. 9.2 (**a**) Network Topology, (**b**) Group Members in Network

members with a frame. We go from each "neighbor" and expand the network to the individuals who directly connect to the "neighbors." This results in an extended egonetwork system.

Networks that are extracted using snowball sampling have the potential problem that the networks may overlap. To some extent, this would affect model calibration. To construct non-overlapped networks, after snowball-sampling computations, we perform a community detection algorithm to obtain different networks which don't overlap. For a massive amount of telecom data, we recommend using the Louvain Method for community detection [28] The Louvain method is efficient in extracting non-overlapping communities from a large network. It operates through two steps: First, it assigns nodes into small communities, favoring local optimization of network modularity. Second, it defines new coarse-grained networks by aggregating nearby communities found in the first step. The two steps iterate until there is no modularity-increasing reassignments of communities.

9.3 Network Measures

Network structure can be measured by multiple graph properties. For example, there are local measures and global measures. Local measures basically are measures of individual positions on the local/ego network, such as individual centrality and the individual clustering coefficient, based on the ego network perspective. On the other hand, measures include density, global clustering coefficient, eigenvalues, and entropies of the individual centrality measures, which are all based on a network, constructed using snowball network strategy or full network strategy. Another way to categorize network structure measures is based on the characteristics of a network being captured by the measure. It can be divided to the measures of number (e.g. degree, density), the measures of centrality (e.g. degree centrality, betweeness centrality and eigenvector centrality), the measures of transitivity (e.g. clustering coefficient) and the measures of variability (e.g. Shannon entropy of the centralities).

Individual local network measures provide information on an individual's position within the network. In Chap. 12, we consider three centrality measures (degree, betweenness, and eigenvector), and the individual clustering coefficient. The degree centrality evaluates an individual's importance in a network by counting how many connections the individual owns. The betweenness centrality shows that an individual is important if the individual always stays on the paths that connect to different others. The eigenvector centrality reflects how important an individual is by who the individual connects to. If the individual connects to important others, then the individual is also important. The individual clustering coefficient quantifies the degree to which an individual's local neighbors form a closure.

Global measures focus on the general structure of the network itself, which includes properties including network size, density, the global clustering coefficient, assortativity, maximum and minimum eigenvalues, and three entropy measures based on the degree, betweenness, and eigenvector centralities. The network size and density questions whether larger and more enhanced network connections (more dense) will result in higher occurrences of social influence. Network size determines the pool of customers that can potentially reach and interact with each other, and network density is the average degree (number of links each individual has) in the network. The global clustering coefficient measures how cohesive or closely knit a network is to the extent that the friends of an individual are also friends with one another. Assortativity measures the extent to which highly connected nodes are more likely to be connected to other high degree nodes in the network according to the node-to-node degree of correlation [27]. The maximum eigenvalue can be used to predict diffusion threshold for the percolation process, and the minimum eigenvalue of a network matrix captures the extent to which the network amplifies the substitutability of an agent. Lastly, centrality entropy measures how evenly distributed are connections between individuals.

9.4   Modeling Consumer Behaviors Within a Network

Depending on what we want to study, type of data collected or chosen methodology, modeling strategies may change accordingly. In Chap. 10, because our interest is in churn rate under social influence, we use a dynamic forward-looking model to capture both the social-learning effect and the network effect in the stay/switch decision of an individual. In Chap. 12, we also use a dynamic model. However, in this study, the question of interest is product diffusion within a network. To examine how network structure and dynamics interplay with the effect if social influence to facilitate diffusion, we extracted individual social networks using snowball sampling based on Call Detailed Records of users. Because we want to separate the effects of social influence and homophily, and at the same time retain the social networks of individuals (rather than considering the whole population as a single network), we use the stochastic actor-based dynamic network model (RSEINA). This model estimates the co-evolution of network formation and adoption behaviors using longitudinal network information, and is able to identify and quantify the social influence effect.

The study in Chap. 11 demonstrates the use of a static model. In Chap. 11, we model individual adoption behavior using the spatial probit model which is adapted from the classic linear spatial autoregressive (SAR) model. In this study, we again use snowball sampling and the Louvain method [28] for network construction. We then use the spatial probit model to control for the effect of network formation, and then estimate the effects of local and global network effects on adoption decisions of individuals.

One thing to notice when constructing social network is the endogeneity issue of network formation. This is one of the factors unique to modeling consumer behavior within a network, where some unobserved individual characteristics that affect consumer behavior may also affect the selection of contacts. These correlations may result in biased estimates for the network related variables in the SAR model, such as peer influence and the network structure effects. We solve this problem by introducing a network formation model following the latent variable approach of Hsieh and Lee [29] to capture the individual unobserved heterogeneity in the network formation process. The central idea of modeling network formation is based on the concept of homophily [30], which states that individuals with similar characteristics are more likely to be connected.

In this case, the spatial probit model and the network formation model are combined as simultaneous equations. We incorporate the unobserved latent variables from the network formation equation into the outcome equation of the spatial probit model to correct for the network endogeneity. The dyad-specific variables between individuals that are used in the network formation equation do not generally affect the behavioral outcomes, and thus, serve as natural exclusion restrictions to identify the simultaneous system. These dyad-specific variables serve as good candidates for exclusive restrictions that are valid in many different application contexts. The proposed model also allows the inclusion of ad hoc exclusive restrictions depending on the research context.

References

1. Iyengar, Raghuram, Asim Ansari, Sunil Gupta (2007), "A model of consumer learning for service quality and usage," Journal of Marketing Research, 44(4) 529–544.
2. Grubb, Michael D (2012), "Dynamic nonlinear pricing: biased expectations, inattention, and bill shock," International Journal of Industrial Organization, 30(3) 287–290.
3. Ascarza, Eva, Anja Lambrecht, Naufel Vilcassim (2012), "When talk is free: The effect of tariff structure on usage under two-and three-part tariffs," Journal of Marketing Research, 49 (6) 882–899.
4. Gopalakrishnan, Arun, Raghuram Iyengar, Robert J Meyer (2014), "Consumer dynamic usage allocation and learning under multipart tariffs," Marketing Science, 34(1) 116–133.
5. Eagle, Nathan, Alex Sandy Pentland, and David Lazer (2009), "Inferring friendship network structure by using mobile phone data." Proceedings of the National Academy of Sciences 106, no. 36: 15274–15278.
6. Onnela, J. P., Saramäki, J., Hyvönen, J., Szabó, G., Lazer, D., Kaski, K., Kertész, J. and Barabási, A.L. (2007). Structure and tie strengths in mobile communication networks. Proceedings of the National Academy of Sciences, 104(18), 7332–7336.
7. Eagle, Nathan, Michael Macy, Rob Claxton (2010), "Network diversity and economic development," Science, 328(5981) 1029–1031.
8. Godinho de Matos, Miguel, Pedro Ferreira, David Krackhardt (2014), "Peer influence in the diffusion of the iphone 3g over a large social network," Management Information Systems Quarterly (Forthcoming).
9. Hu, Mantian, Sha Yang and Daniel Yi Xu (2016), "Learning and Network Effects within Social Networks," working paper, CUHK, Hong Kong.
10. Mantian Hu, Chih-sheng Hsieh, and Jamie Jia (2015), "Predicting Peer Influence Using Network Structure," working paper, CUHK, Hong Kong.
11. Mantian Hu, Chih-sheng Hsieh, and Jamie Jia (2016), "Network Based Targeting: The Effectiveness of Peer Influence within Social Networks," working paper, CUHK, Hong Kong.
12. Gonzalez, M. C., Hidalgo, C. A., & Barabasi, A. L. (2008). Understanding individual human mobility patterns. Nature, 453(7196), 779–782.

13. Song, C., Qu, Z., Blumm, N., & Barabási, A. L. (2010). Limits of predictability in human mobility. Science, 327(5968), 1018–1021.
14. Palchykov, V., Mitrović, M., Jo, H. H., Saramäki, J., & Pan, R. K. (2014). Inferring human mobility using communication patterns. Scientific reports, 4.
15. Moretti, Enrico (2011), "Social learning and peer effects in consumption: Evidence from movie sales." The Review of Economic Studies 78, no. 1: 356–393.
16. Chandrasekhar, Arun, Horacio Larreguy, and Juan Pablo Xandri (2012), "Testing models of social learning on networks: Evidence from a framed field experiment." Work. Pap., Mass. Inst. Technol., Cambridge, MA
17. Zhang, Juanjuan (2010), "The sound of silence: Observational learning in the US kidney market." Marketing Science 29, no. 2: 315–335.
18. Cai, Hongbin, Yuyu Chen, and Hanming Fang (2009), "Observational learning: Evidence from a randomized natural field experiment." American Economic Review 99, no. 3: 864–882.
19. Ellison, Glenn, and Drew Fudenberg (1995), "Word-of-mouth communication and social learning." The Quarterly Journal of Economics (1995): 93–125.
20. Golub, Benjamin, and Matthew O. Jackson (2010), "Naive learning in social networks and the wisdom of crowds." American Economic Journal: Microeconomics 2(1), 112–149.
21. Bala, Venkatesh, and Sanjeev Goyal (1998), "Learning from neighbours." The Review of Economic Studies 65, no. 3: 595–621.
22. Zhang, Juanjuan, and Peng Liu (2012), "Rational herding in microloan markets." Management Science 58, no. 5: 892–912.
23. Katz, Michael, and Carl Shapiro (1985), "Network externalities, competition and compatibility," The American Economic Review, 75, no. 3: 424–440.
24. Goolsbee, Austan, and Peter Klenow (2002), "Evidence on learning and network externalities in the diffusion of home computers," Journal of Law and Economics, 45, no. 2: 317–343.
25. Young, H. P. (2009). Innovation diffusion in heterogeneous populations: Contagion, social influence, and social learning. The American economic review, 99(5), 1899–1924.
26. Chen, X., Chen, Y., and Xiao, P. (2013). The impact of sampling and network topology on the estimation of social intercorrelations. Journal of Marketing Research, 50(1), 95–110.
27. Kiss, I. Z., Green, D. M., and Kao, R. R. (2008). The effect of network mixing patterns on epidemic dynamics and the efficacy of disease contact tracing. Journal of The Royal Society Interface, 5(24), 791–799.
28. Blondel, Vincent D, Jean-Loup Guillaume, Renaud Lambiotte, Etienne Lefebvre. 2008. Fast unfolding of communities in large networks. Journal of statistical mechanics: theory and experiment 2008(10) P10008.
29. Hsieh, Chih-Sheng, Lung-Fei Lee. 2016. A social interactions model with endogenous friendship formation and selectivity. Journal of Applied Econometrics 31(2) 301–319.
30. Lazarsfeld, Paul F, Robert K Merton, et al. 1954. Friendship as a social process: A substantive and methodological analysis. Freedom and control in modern society 18(1) 18–66.

Chapter 10
Contagious Churn

10.1 Introducing the Problem

For highly competitive markets such as mobile carriers, one of the main concerns is to preserve and expand their customer base. Developing marketing strategies to expand total number of customers is a top priority as it is their main source of revenue. It can also act as a representative of the brand, attracting new customers via their social networks and preventing peers from switching operators. Mobile carriers may choose to either bring new users by increasing attractiveness of their brand, or reduce switching customers by improving service quality, and identifying and managing potential leaving clients. These two commonly used strategies to managing a customer base are also combinable. However this chapter will only be focusing on the latter.

In most developed countries, China included, the telecom market is close to saturation level, such that most individuals already own one or more cell phone. New customers will be hard to identify and they will be reluctant to switch to a new mobile carrier for no particular reason. It is often an issue in these countries where few new customers can be reached, and the best way to expand their customer base and attract new users is to convince them to switch from a current carrier. This focus on customer retention is also known as the churn rate problem.

10.1.1 Churn Rate Problem

The churn rate problem is important in telecom marketing and saturated markets because retention of customers is much less costly than acquiring new customers. Since most potential customers from other carriers are already satisfied with the services and prices, they will not switch unless there is a large motivating factor. One possible factor is peer influence, where customers influence each other to switch

© Springer International Publishing AG, part of Springer Nature 2018
Y. Ouyang et al., *Mining Over Air: Wireless Communication Networks Analytics*,
https://doi.org/10.1007/978-3-319-92312-3_10

carriers. Studying the churn rate problem is also critical for mobile carriers to maintain sustainable growth. Using the vast amount of information mobile carriers can receive from their customers; they can understand customer behavior and design the best strategy to retain customers.

In this chapter, we will analyze switching behavior of customers from one carrier to another to predict customer churn rate. We will propose a model to integrate both social-learning and network effects into a dynamic model, which allows us to compare effects of the two components. In this study, our goal is to go beyond simply documenting this phenomenon to explain why and how the phenomenon of contagious switching behavior occurs, by applying an analytic methodology based on assumptions coming from social science and econometrics.

10.1.2   Social Learning and Network Effects

One of the reasons that contribute to customers switching carriers is peer influence; this phenomenon also known as contagious switching. According to a survey conducted by ComScore Networks in 2007, when U.S. wireless phone subscribers were asked about their behaviors and attitudes on their usage, 13% of respondents considered the influence of friends and family as their primary reason for switching carriers. Contagious switching is a network-related behavior and states that consumers are more likely to switch carriers if more of their contacts from the same carrier had switched. Recently, this phenomenon has attracted some amount of attention in the telecommunications industry, including the work by Richter et al. [2]. Examples from many other fields also provide example of instances where other consumers' behavior have influence over an individual's decision on the same behavior [3–5]. The same effect have been observed in marketing, particularly in the context of adoption of new product or new technology [6–11].

There are two fundamentally different explanations that may explain why individuals tend to imitate others' behaviors [12]. The first explanation is referred to as social learning, and is based on the information exchange argument, which suggests that individuals use feedback from peers to update their own expectations of product quality [13]. In social learning, there are two separate communication channels: direct and indirect. An example of the direct channel in social learning is word-of-mouth communication [14], where peers exchange information directly. Then, individuals may indirectly receive information from their peers through observational learning [15, 16], where observation of others' behaviour to infer their peer may have some prior knowledge that one does not process. For this model, we assume that people gain information through both channels and aggregate the different sources of information.

The second explanation for why people imitate others' behavior is based on the direct benefit gained from the network. People will follow others if direct utility can be gained by aligning their behavior with that of others. This is the network effect, which refers to particular forms of externalities such that the actions of peers (i.e., the network size) directly affect an individual's preference [17]. Instead of indirectly updating their information set like in social learning, individuals act because others' behavior directly affects one's payoff.

For example, take two wireless carriers A and B. Carrier B enables a preloaded video chat app over the mobile network for all customers, whereas carrier A only allows video calls over Wi-Fi. When a person's friends all switch to carrier B from carrier A, he or she is very likely to switch as well. The underlying motive behind switching carriers could be because one's peers present new information about an alternative carrier and the person uses that information to update his or her own perception of carrier B's service (social-learning effects), or simply because it is easier and cheaper to video chat if they all have the same carrier (network effects). In the latter case, even after the individual had updated his or her perceptions about carrier B, and determines that carrier B may not be as good as carrier A in other aspects, he or she may still want to switch due to the direct payoff of the ease of communicating with peers.

From a managerial perspective, it is important to disentangle the social-learning effect and the network effect, because the strategic implications from the two effects are very different. If social learning turns out to be the main driver for a user's switching behavior, the managerial implication is to provide better products and to stimulate positive word-of-mouth among users to prevent customers from switching. However, if the network effect plays a dominant role, the company would be better off maintaining a large customer base. In such a case, strategies that include price discrimination against phone calls made to outside networks or exclusive offers of apps and games to its own subscribers would benefit the company.

In the recent years, both social-learning and network effects have been studied extensively. For example, Ching [18] developed an empirical structural demand model for prescription drugs in which he modeled the process of information aggregation based on consumers' experience to update a common belief about a new product's quality over time. Narayan et al. [19] studied the manner in which peer influence affect the choices of products with multiple attributes and found that consumers update their attribute preferences by integrating their prior beliefs with the preferences of others in a Bayesian manner. Also focusing on consumer learning on multiple product attributes, Chan et al. [20] studied the manner in which physicians learn about treatment effectiveness and side effects from detailing and patient feedback. Chintagunta et al. [21] also incorporated learning from patients' feedback to model a physician's prescription choices. Zhao et al. [22] proposed a Bayesian learning model to capture the ways that consumers learn from their own experience with one genre of literature and other reviewers' experience with a particular book to form an overall expectation on the quality of that book. They also found that the Bayesian learning model outperforms a reduced-form model with only state-dependence effects from previous experience. For more details, Jackson [23] and Mobius and Rosenblat [24] had provided comprehensive reviews on social learning in networks in the field of economics.

There have also been considerable amount of progress in studying networks since the formal development of a model of network effects by Katz and Shapiro [17].

Goldenberg et al. [25] had provided an excellent literature review on this topic. In the context of this chapter, we will focus on the local/direct network effect in which an individual makes decisions in relation to his or her close social network. Tucker [26] used exogenous shocks to identify the network effect of one individual on another's decision regarding new technology adoption and to identify more influential individuals. Ryan and Tucker [27] developed a model to study the demand for video calling technology that took into consideration both network effects and heterogeneity; they found that people have different adoption costs and network benefits and that they prefer diverse networks.

Despite the vast amount of literature on social-learning and network effects, very few papers have studied the mechanisms together. Goolsbee and Klenow [28] were among the first to claim that network externalities and learning from others can be treated as local spillovers and that they affect the diffusion of home computers so that people are more likely to purchase their first home computer if a large share of their peers own computers. However, rather than distinguishing between the two mechanisms, they focused on the identification of local spillover effects from the unobserved common traits of households. For example, using aggregate movie box office and weather data, Moretti [13] had empirically identified social learning from network externalities and found social learning to be an important determinant of movie ticket sales.

In this study, we were able to use data from one wireless mobile carrier to study the individual stay/switch decision, by combining the two underlying behavioural mechanisms in a single model and to distinguish between their effects. For this purpose, we used individual data with information about behaviour and network change. To explain the individual consumers' stay/switch decisions, we focus our attention on three separate mechanisms. First is the learning-from-self mechanism. In Iyengar et al. [29], it states that the service quality of the current carrier will directly affect the utility of staying. The customer faces uncertainty due to service variability between and within mobile carriers, and learns about the mean quality of the carrier they are subscribed to from their own usage experience.

The second and third mechanisms are social learning and the network effect, as previously described. These two mechanisms capture the manner in which others' switching behavior affects an individual's switching decision. In social learning, the quality expectation of alternative carrier, with such knowledge acquired from peers, will affect the customer's switching probability. This knowledge may come from observation of the actions taken by others or from direct communication with those in their contact network who have switched, and result in an update of the individual's own product quality expectations of alternative carriers.

In this context, our model accounts for two unique aspects of consumers' strategic learning from others: (i) the signal on an alternative carrier from a contact who has switched is perceived to be systematically different by the individual, depending on whether the signal comes from a loyal contact who switched from the focal carrier. Consequently, loyalty of peers who switched from the focal carrier may systematically affect the perception of their evaluation of the alternative carrier and that the information from close friends is perceived as less noisy; and (ii) the noisiness of the signal on an alternative carrier from a contact who has switched is dependent on the strength of the relationship between the individual and the contact. Finally, in network effect, the number of the individual's contacts who subscribe to the same mobile carrier will directly affect the consumer utility of staying.

10.2 Dealing with Network Data

In this study, data were obtained for one country from a third-party European research firm that specializes in the multinational telecommunications industry. The data collection period spanned from June 2008 to February 2009. Data of interest was individuals' calling behavior within one mobile carrier. Eagle et al. [30] compared mobile network data and self-reported survey data and found that "the data collected from a mobile network can be used to accurately predict cognitive constructs such as friendship." This finding suggests that mobile network information would be useful in studying other consumers' influence on the stay/switch decision with respect to subscription to a wireless carrier.

Two separate sets of data have been collected. The first set includes calling record data and the aggregated information of each individual's mobile service usage (including the number of phone calls, text messages sent and received, and their contacts' information) over a short period of time at the beginning of the sampling period. This dataset was used to identify each customer's initial network. Then, the second dataset contains personal information from each individual subscriber, such as demographics, plan details (postpaid or prepaid, on-network or off-network cost ratio, minutes included, and family plan ID), and churn dates, that were collected every month from June 2008 to February 2009.

Here, we are interested in studying the interaction between users. To define the boundary of networks and calculate related variables, two separate algorithms have been used. First, snowball sampling is computed to extract network of friends from the entire database. We use snowball-sampling because it performs better in preserving the network structure [31]. 20 individuals ("seeds") were randomly selected from the entire mobile network. We then identified everyone the seeds were in contact with and labelled them as neighbors or "friends" of the seeds. Together, they formed a sphere of a network group. Then we form a second layer using information about the direct contacts of the neighbor's neighbors. We thus formed a two-layer snowball for each seed.

Since snowball networks are potentially overlapped and cannot warrant a clean environment to conduct cross-network analysis, a community detection method is then performed. In total, 198 networks have been identified, including 2077 individuals. Complete network and demographic information for each individual is included in the sample. After the snowball-sampling computations, we obtain several different networks which can be overlapped, and we perform a community detection algorithm. We follow the Louvain Method for community detection as demonstrated by Blondel et al. [32] to extract the non-overlapping communities from, the large network. Some other options exist to detect communities. One alternative sampling and network building algorithm is the modified T-CLAP [33], which combines snowball sampling and network clustering procedures. However, the modified T-CLAP do not prevent overlapping of the identified networks. Another well-known method is the iterative edge-removal approach in Newman and Girvan [34]. This method continuously finds out and removes the edge which owns the largest betweenness measure to separate a network. The main advantage for the Louvain method is its fast computing speed: It appears to run in time $O(n \log n)$, where n denote the network size, so any graphs with up to 109 edges can be analyzed in a reasonable time on current computational resources.

10.3   Dynamic Model

10.3.1   Introduction of the Model

We propose a dynamic forward-looking model that captures both the social-learning effect and the network effect in the context of an individual's decision on whether to stay or switch with respect to a wireless carrier. The model consists of three key elements. First, each consumer learns about the quality of the current wireless carrier through his or her own experience. Mobile service is an experience good; customers are usually uncertain about its quality before they actually use it, and it is not enough for them to use it just once. They form a general perception about the quality through repeated usage over time. The perceived quality of the same wireless carrier can vary across customers, potentially due to network coverage or the efficiency of the devices. Customers usually have a prior belief regarding the quality, and we allow them to update this belief in a Bayesian fashion [35].

Second, because learning from others about alternative wireless carriers is much less costly than sampling or obtaining direct experience, we assume that people use feedback from their peers to update their own expectations of product quality for these alternative wireless carriers. We infer the occurrence of this social learning when a person's mobile contacts change carriers. Their prior beliefs about the alternatives' service quality vary depending on the customer's idiosyncratic beliefs. Updating occurs when a member of the person's contact network switches carriers. We assume that individuals update their beliefs about the alternative carrier in a Bayesian fashion when a member of his or her network switches carriers. To capture consumers' strategic learning from others, we allow the signals from others on alternative carriers to be perceived to be systematically different depending on the individual consumer's loyalty status, and we allow the credibility of the signals from others to be dependent on the closeness of the relationship between the individual and the contact who recently switched from the focal carrier.

Third, we further assume that the utility from using the service depends directly on the number of contacts in the network. The network size may enter a utility function by three mechanisms [13]. The first is customers' cost considerations. For example, some wireless carriers offer plans with free text messages to people on the same network. Thus, if more friends switch networks, higher costs may be incurred later due to the absence of free text messages. Second, as smart phones become more popular, mobile social applications may allow friends to play games on their mobile phones regardless of the time or location. Therefore, having more friends on the same network with the same game may lead to a greater amount of fun that the customers get out of the experience. The third mechanism is referred to as "conformism" by sociologists. People tend to behave the same as their peers in the same group. The strength of this effect is usually modeled to be affected by the number of people who make the same choice in the group (i.e., the network size).

We assume that consumers are forward-looking. Using a questionnaire and experiment, Lemon et al. [36] showed that "when deciding whether or not to continue a service relationship, consumers not only consider current and post evaluations of the firm's performance, but also incorporate future considerations regarding the services." This forward-looking behavior also works through two anticipations: the anticipated future benefit and the anticipated future regret. Because the network size directly affects customer utility, it results in an intertemporal link when customers make their decisions. Consumers who decide to switch today will take into consideration the extra cost that they may incur tomorrow because their network is different from that of their friends. In addition, uncertainty about quality induces a tradeoff between switching today and bearing all of the risks versus waiting and accumulating more information.

10.3.2 Definition of the Model

Following the dynamic structural literature (e.g., [37, 38]), we summarize the conditions of each individual i at each time with $I_{i, t}$ a vector of state variables. In the following, i is not mentioned since it is the user of interest which is fixed: $I_{i, t} = I_t$. This vector contains the whole information for the individual at date t: General state variables, received quality signals from the current and the alternative network and random shocks which influence user choice to switch or to stay.

The model developed stands in the game-theory context, where each user do a rational choice by trying to maximize his utility function to decide whether to switch or not. The shape of the equations follows Bellman's equations, and is solved using programming algorithms. The following gives an explicit description of the model and of the different components. This part is more technical and can be skipped in the first reading.

Explicitly, general state variables are $z_t = (Net_t, X_t)$, where Net_t is the network size and $X_t = (Price, Min, Fam, Male, Age)$ includes mobile plan information and individual characteristics: Relative on-network/off-network call price of the current plan for each individual, which is predetermined and does not vary across time; Number of minutes included in the plan; Binary variable in which 1 indicates that the customer is subscribed to a family plan and 0 otherwise; Gender and; Age. Received quality signal of the focal network is S_{ft}, and received quality signal estimated by the customer for the alternative network is S_{at}. Before taking action, each individual receives a private choice-specific preference shock to switch $\varepsilon_t(0)$ and to stay $\varepsilon_t(1)$ that is known to the individual but is unobserved by the researchers. By assumption, it is drawn independently across agents and over time. Random shocks will influence user choice at each time.

Given the state It, consumers predict their own network size changes and make switching decisions at the beginning of each period. If they decide to switch, their contract terminates by the end of the period. Let $d_{it} \in \{0, 1\}$ denote individual i's decision at time t. This is a dichotomous indicator that equals 1 if consumer i choose to stay with the network in period t and 0 otherwise.

We first summarize the different parameters to be estimated or fixed in the model. $\beta \in (0, 1)$ is the common discount factor and is set as 0.99. We further assume that when customers switch, they do not return and there is no new entry into the network because we do not observe a new entry in our sample. General state variables X_t are weighted linearly using real parameters $\theta = (\theta_1, \ldots, \theta_5)$. Power of local network effect is related to a parameter λ. Finally, r measures the risk aversion.

The consumers' single-period utility from staying with their current wireless carrier in time period t, u_{1t} is assumed to be a function of z_t and S_{ft}. Following Crawford and Shum [38], we consider a quasilinear utility specification with a constant absolute risk-aversion specification for the subutility function of S_{ft} and linear terms in z_t,

$$u_{1t}(z_t, S_{ft}; \theta) = -\exp(-rS_{ft}) + \theta\, X_t + \lambda Net_t, \tag{10.1}$$

Furthermore, we define $u_{0t}(S_{at})$ as the abstract utility function for the alternative wireless carriers, which is only a function of S_{at}:

$$u_{0t}(S_{at}; \theta) = -\exp(-rS_{at}). \tag{10.2}$$

Before including discounted values and random chocks to change, we can notice that when $r >> 1$, risk aversion is high and $u_{1t}(z_t, S_{ct}; \theta) = \theta\, X_t + \lambda Net_t$ whereas $u_{0t}(S_{at}; \theta) = 0$, thus individual cannot switch. Also, when $0 < r << 1$, individual takes all risks, $u_{1t}(z_t, S_{ct}; \theta) = -1 + rS_{ct} + \theta\, X_t + \lambda Net_t$ and $u_{0t}(S_{at}; \theta) = -1 + rS_{at}$. Since in this case effect of X_t and Net_t are small [38], there is only a competition between S_{ct} and S_{at} to switch, without other effects.

Consumers are assumed to switch carriers if the current network utility plus the discounted switching value and the idiosyncratic taste of switching (called utilitySwitch(t) in the following) is greater than the current utility plus the discounted continuation value and the idiosyncratic taste of continuing (called utilityStay(t) in the following). The Bellman equation for the value of people who are currently on the network at the beginning of each period is taken from Pakes et al. [37] and Dunne et al. [39]. We can define the one-step utility, which corresponds to check gain to switch or not at date $t + 1$, without taking into account time after $t + 1$. From previous notations, we have:

$$V_1(t) = u_{1t}(z_t, S_{ct}; \theta) + \beta VC(I_t; \theta) + \varepsilon_t(1), \tag{10.3}$$

$$V_0(t) = u_{1t}(z_t, S_{ct}; \theta) + \beta VX(I_t; \theta) + \varepsilon_t(0). \tag{10.4}$$

We take the "current utility" at the both side since the individual is in the current carrier at date t, and at that time decides to switch or not in the future, where the gain for each is represented with "discounted switching value" and "discounted continuation value".

Then, we define VC(.) the continuation value, which is the expectation of the next period's realization for the whole future of the value function (mathematically, this corresponds to integration on future possibilities to switch or not depending of the future utilities functions):

$$VC(I_t; \theta) = E[\max(\text{utilityStay}(t+1), \text{utilitySwitch}(t+1)) \mid I_t] \tag{10.5}$$

With:

$$\text{utility Stay}(t+1) = u_{1t+1}(z_{t+1}, S_{ct+1}; \theta) + \varepsilon_{t+1}(1)$$
$$+ \beta \int_{I'_{t+2}} VC(I'_{t+2}; \theta) dP(I'_{t+2} | I_{t+1}),$$
$$\text{utility Switch}(t+1) = u_{1t+1}(z_{t+1}, S_{ct+1}; \theta) + \varepsilon_{t+1}(0)$$
$$+ \beta \int_{I'_{t+2}} VX(I'_{t+2}; \theta) dP(I'_{t+2} | I_{t+1}).$$

Conversely, VX(.) is the expected utility from the alternative wireless carriers, which represents the switching value. In the model, since we have no information about other carriers and did not observe return of a switching user, we consider that individuals cannot switch anymore when they already switched once. The formula is then simpler:

$$VX(I_t; \theta) = E[u_{0t+1}(S_{at+1}; \theta) \mid I_t]. \tag{10.6}$$

In the language of dynamic programming, those functions VC and VX are called choice specific value functions. They will be the key to solve the dynamic problem. At this point, the shape of the model is defined iteratively. To be able to explicitly resolve the model, we need to define how utility function are related to own experience, social learning and network effect, by defining S_{ft}, S_{at} and by estimating the network at time $t + 1$.

10.3.3 Modeling Own Experience, Social Learning and Network Effects

Although consumers are uncertain about the quality of experience goods, they are assumed to have prior information and to learn each time from own experience. Each consumer's prior beliefs about the perceived true quality, Q_{if}, is summarized by a normal distribution:

$$Q_{if} \sim N\left(\mu_f, \sigma_{\mu f}^2\right). \tag{10.7}$$

In this expression, μ_f denotes the prior mean quality and the standard deviation, $\sigma_{\mu f}^2$, measures the precision of the prior means. Consumer i does not know Q_{if} but receives quality signals that allow him or her to update his or her perceptions about the true quality. We assume that the signals are independently and normally distributed across periods around the true service quality (Q_{if}):

$$S_{ift} \sim N\left(Q_{if}, \sigma_f^2\right). \tag{10.8}$$

S_{ift} is observed by the consumer but is unknown to the researchers, whereas Q_{if} is unknown to either. The usage experience and subsequent observed signal S_{ift} do not exactly reveal Q_{if} but do provide information about it. This distribution assumption, along with an initial conjugate prior on Q_{if}, generates a Bayesian learning process in which the consumer's posterior beliefs about Q_{if} are given recursively (see [40]). Explicitly, we build a sequence of normal distributions with means π_{ift} and variance V_{ift} satisfying the following Eqs. (10.9) and (10.10), where nit + 1 is the total number of periods that consumer i stays with the current carrier until period t + 1. We also recall that $d_{it} \in \{0, 1\}$ denotes individual i's decision at time t.

$$\pi_{ift+1} = \begin{cases} \dfrac{\dfrac{\pi_{ift}}{V_{ift}} + \dfrac{S_{ift+1}}{\sigma_f^2}}{\dfrac{1}{V_{ift}} + \dfrac{1}{\sigma_f^2}} = \dfrac{\sigma_f^2}{V_{ift} + \sigma_f^2}\pi_{ift} + \dfrac{V_{ift}}{V_{ift} + \sigma_f^2}S_{ift+1} & \text{if } d_{it} = 1 \\ \pi_{ift} & \text{otherwise} \end{cases} \tag{10.9}$$

$$V_{ift+1} = \begin{cases} \dfrac{1}{\dfrac{1}{\sigma_{\mu f}^2} + \dfrac{n_{it+1}}{\sigma_f^2}} & \text{if } d_{it} = 1 \\ V_{ift} & \text{otherwise} \end{cases}, \tag{10.10}$$

Furthermore, we assume that the choice-specific preference shocks $\varepsilon_{it}(d_{it})$ have an extreme value distribution. The conditional choice probability then has a very convenient expression:

$$P(d_{it}|I_t) = \frac{\exp(V_{d_{it}}(I_t))}{\exp(V_1(I_t)) + \exp(V_0(I_t))} \tag{10.11}$$

Customers also learn from other. It is less costly and more efficient than sampling or obtaining direct experience because subscribing to and learning about a new wireless carrier is often costly. When a contact on the network switches, it not only affects the size of the network but also helps the individual update the quality information about alternative wireless carriers. Q_{ia} Denotes the perceived true quality of alternative wireless carriers for individual i. We normalize the prior mean for alternative wireless carriers to zero and assume the equality of initial variances, for the purposes of empirical identification:

$$Q_{ia} \sim N\left(0, \sigma_{\mu f}^2\right). \tag{10.12}$$

Learning occurs only when a friend on the consumer's network switches, so there may be no learning at all for some periods or multiple learning occurrences for a particular period. Furthermore, we assume that the social-learning process is strategic in two ways. First, the signals on alternative carriers from contacts who have switched may be perceived differently by the individual, depending on the type of contact (e.g., loyalty). Intuitively, if the contact who switched had been a highly loyal customer with the focal carrier, the individual may interpret the signal from this contact to be more positive than the mean. This is because alternative carriers often need to stimulate a better-than-average experience for a switcher who was loyal to a competitor's service. To reflect this logic, we specify the mean of the signals on alternative carriers to equal $Q_{ia} + \delta z_j$, where zj takes the value of 1 if contact j had been a loyal customer of the focal carrier (i.e., they renewed their contract) and 0 otherwise. Following our rationale, the value of δ would be positive.

Second, we anticipate that consumers may perceive signals from other people to carry different levels of credibility. Intuitively, signals from closer contacts are less noisy than signals from more distant contacts. To reflect this logic, we allow the variability of the experience signal to be a function of the distance between the individual consumer and the contact who switched, $\exp\left(\alpha_0 + \frac{\alpha_1}{D_{ij}}\right)$. D_{ij} measures the contacting frequency between the contact j and the individual i (i.e., $1/D_{ij}$ measures the distance between the two). Therefore, the value of α_1 is expected to be positive. Econometrically, the use of those two aspects allows us to account for the individual heterogeneity of the contacts and the dyadic heterogeneity between the individual and his or her contacts.

To reflect such a strategic social-learning behavior, we specify quality signals on alternative carriers as follows:

$$S_{ia}^t \sim N\left(Q_{ia} + \delta z_j, \exp\left(\alpha_0 + \frac{\alpha_1}{D_{ij}}\right)\right) \tag{10.13}$$

As for learning from own experience – Eqs. (10.9) and (10.10), this generates a recursive Bayesian learning process.

Finally, to model network effect, we follow Pakes et al. [37], and we let Net_{it} be the size of individual i's network at the beginning of each period and let NSW_{it} be the number of contacts who switch during time period t (which is unknown at the time of the consumer's decisions). To calculate VC(.) from model, we need to form the consumer's perceptions about the likely numbers of contacts who switch conditional on the current state variables. These perceptions generate the probability distribution $P(NSW_{it} | Net_{it}, X_{it})$. It is required that $P(NSW_{it} | Net_{it}, X_{it})$ be consistent with the contacts' behavior so as to generate equilibrium; In other words, consumers have rational expectations. We let

$$Net_{it}^e = Net_{it} - NSW_{it}^e = Net_{it} - \int NSW_{it} dP(NSW_{it} | Net_{it}, X_{it}), \qquad (10.14)$$

where Net_{it}^e and NSW_{it}^e denote the expectations conditional on the current state.

10.3.4   Estimation of the Model-Alex

We solve the model with a two-step dynamic programming algorithm, with the assumption that a consumer is forward-looking and decides whether to stay with the same service carrier in each period by maximizing the total utility received from that day onward. Direct implementation of the method used by Bajari et al. [41] in our context is not feasible because the service qualities are unobserved, which prevents us from estimating the transition probabilities and the policy functions for each individual. Therefore, our approach uses the idea of two-step models [37, 41, 42] but embeds the first step in a nested fixed-point algorithm. In other words, instead of estimating who switches in the first step, we estimate how many switch. With our notations, we first try to estimate NSW_{it}^e for each individual from its formula. Therefore, knowing NSW_{it}^e which is now assumed known, we calculate the network size at the next time. Simulation of the signals S_{ift} and S_{iat} is possible.

In equilibrium, the perceptions of network size change must be consistent with actual observations. Our proposed estimation consists of two steps. In the first step, we estimate NSW_{it}^e for each individual. From data exploration, we assume that NSW_{it}^e can be approximated by a multinomial ($NSW_{it}^e = 0, 1, 2$) logit model of Net_{it} and X_{it}. Because the consumers' expectations must be consistent with the average realizations, these estimations will converge to the true expected number of switching neighbors. The second step treats the estimate of NSW_{it}^e as known and calculates the network size by subtracting this number from the network size at the beginning of the period.

We then estimate the structural parameters using an adaption of the approximation method used by Keane and Wolpin [43] ([38]). We use simulation to integrate the two signals S_{ift} and S_{iat} because they are unobservable to researchers. Thus, the simulated likelihood is as follows,

Since S_{ift} and S_{iat} are unobservable to researcher, we use simulation to integrate the two signals by simulating likelihood. According to (10.11), and with I the number of individuals and S the number of simulations, we obtain:

$$\prod_{i=1}^{I}\left(\frac{1}{S}\sum_{s=1}^{S}\prod_{t}P\left(d_{it}|I_{it}^{s}\right)\right) = \prod_{i=1}^{I}\left(\frac{1}{S}\sum_{s=1}^{S}\prod_{t}\frac{\exp(V_{d_{it}}(I_{t}))}{\exp(V_{1}(I_{t})) + \exp(V_{0}(I_{t}))}\right)$$

(10.15)

We use $S = 7500$ draws for the simulation and the Berndt-Hall-Hall-Hausman method to approximate the Hessian matrix. Finally, the whole parameters can be estimated.

10.4   Results

Table 10.1 reports estimates of parameters of the proposed dynamic structural model. The point estimates for the parameters of the true quality distribution indicate substantial heterogeneity in service quality across customers.

Notice that μf actually captures the average quality difference between the current wireless carrier and the alternatives. The relative mean service quality estimate (0.242; s.e. 0.020) indicates that the focal wireless carrier has a higher average service quality than the alternative carriers. The estimate of the standard deviation

Table 10.1 Model comparison

	Proposed model (standard error s.e.)
Relative price θ_1	−0.506** (0.295)
Plan minute θ_2	0.164** (0.032)
Family θ_3	0.118 (0.152)
Male θ_4	0.572** (0.158)
Age θ_5	0.300** (0.056)
Network λ	0.080** (0.034)
μ_f	0.242** (0.020)
$\sigma_{\mu f}$	1.154** (0.067)
σ_f	1.141** (0.019)
α_1 in S_{ia}^t variance	−1.019** (0.302)
α_2 in S_{ia}^t variance	0.365** (0.115)
Risk aversion r	0.977** (0.032)
Loyalty δ	1.052** (0.013)
Sample size	1885
LL	−223.998
BIC	546.038

**is the omitted precision

of the true service quality, $\sigma\mu f$ (1.154; s.e. 0.067), is large, which suggests a rather large variation in service quality. The standard deviations of the quality signals, σf (1.141; s.e. 0.019), measures how quickly consumer uncertainties about true service quality decreases due to learning from their own usage experience.

The estimate of δ is significantly positive (1.052; s.e. 0.013), indicating that the signals on alternative carriers from a more loyal contact who has switched are interpreted in a more positive manner than are those from a less loyal customer of the focal carrier. The significant and positive sign of $\alpha 1$ (0.365; s.e. 0.115) demonstrate that a closer relationship between the switched contact and the individual leads to the reception of a less noisy signal.

The results obtained are intuitive. The individual's perception of the information from her contacts depends on the characteristics of the contact and the relationship with the contact. If the contact used to be loyal to the focal carrier but eventually switched, the information regarding the alternative carrier from this contact would be perceived as more positive. The consumer may suspect that the contact must have very positive knowledge of the alternative carrier that she doesn't so the contact decides to switch. And it is also quite natural that people will trust the information from her close contacts more than otherwise. So the learning process differs depending on the information source and our results demonstrate that.

A negative estimate of the on-network/off-network price (-0.506; s.e. 0.295; p-value $= 0.077$) indicates that a larger price difference (i.e., a smaller number) leads to a greater likelihood that the customer will stay. In addition, based on the estimation results, the inclusion of more minutes and family plans are both effective strategies to retain customers. The estimate of network effects λ (0.080; s.e. 0.034) is positive and significant, which indicates that in addition to social learning, the actions of friends directly affect a customer's probability of staying.

References

1. More than One-Quarter of Wireless Subscribers Switched to Their Current Carrier to Gain Better Network Coverage. (January 16, 2007). Retrieved from https://www.comscore.com/Insights/Press-Releases/2007/01/Wireless-Subscribers-Switch-Carriers.
2. Richter, Yossi, Elad Yom-Tov, and Noam Slonim (2010), "Predicting customer churn in mobile networks through analysis of social groups." SDM 732–741.
3. Christakis, Nicholas A, James H Fowler. 2007. The spread of obesity in a large social network over 32 years. New England journal of medicine 357(4) 370–379.
4. Dasgupta, Koustuv, Rahul Singh, Balaji Viswanathan, et al. (2008), "Social ties and their relevance to churn in mobile telecom networks." EDBT'08: Proceedings of the 11th International Conference on Extending Database Technology, 668–677.
5. Aral, Sinan, and Dylan Walker (2014), "Tie strength, embeddedness, and social influence: A large-scale networked experiment." Management Science 60, no. 6: 1352–1370.
6. Van den Bulte, Christophe, and Gary L. Lilien (2001), "Medical innovation revisited: Social contagion versus marketing effort." American Journal of Sociology 106, no. 5: 1409–1435.

7. Van den Bulte, Christophe, and Stefan Stremersch (2004), "Social contagion and income heterogeneity in new product diffusion: A meta-analytic test." Marketing Science 23, no. 4: 530–544.

8. Van den Bulte, Christophe, and Yogesh V. Joshi (2007), "New product diffusion with influentials and imitators." Marketing Science 26, no. 3: 400–421.

9. Manchanda, Puneet, Ying Xie, and Nara Youn (2008), "The role of targeted communication and contagion in product adoption." Marketing Science, 27, no. 6: 961–976.

10. Iyengar, Raghuram, Christophe Van den Bulte, and Thomas W. Valente (2011), "Opinion leadership and social contagion in new product diffusion." Marketing Science 30, no. 2: 195–212.

11. Iyengar, Raghuram, Christophe Van den Bulte, and Jae Young Lee (2015), "Social contagion in new product trial and repeat." Marketing Science 34, no. 3: 408–429.

12. Easley, David, and Jon Kleinberg (2010), Networks, crowds, and markets: Reasoning about a highly connected world. Cambridge, U.K.: Cambridge University Press.

13. Moretti, Enrico (2011), "Social learning and peer effects in consumption: Evidence from movie sales." The Review of Economic Studies 78, no. 1: 356–393.

14. Chandrasekhar, Arun, Horacio Larreguy, and Juan Pablo Xandri (2012), "Testing models of social learning on networks: Evidence from a framed field experiment." Work. Pap., Mass. Inst. Technol., Cambridge, MA.

15. Zhang, Juanjuan (2010), "The sound of silence: Observational learning in the US kidney market." Marketing Science 29, no. 2: 315–335.

16. Cai, Hongbin, Yuyu Chen, and Hanming Fang (2009), "Observational learning: Evidence from a randomized natural field experiment." American Economic Review 99, no. 3: 864–882.

17. Katz, Michael, and Carl Shapiro (1985), "Network externalities, competition and compatibility," The American Economic Review, 75, no. 3: 424–440.

18. Ching, Andrew T. (2010), "Consumer learning and heterogeneity: Dynamics of demand for prescription drugs after patent expiration." International Journal of Industrial Organization 28, no. 6: 619–638.

19. Narayan, Vishal, Vithala R. Rao, and Carolyne Saunders (2011), "How peer influence affects attribute preferences: a Bayesian updating mechanism." Marketing Science 30, no. 2: 368–384.

20. Chan, Tat, Chakravarthi Narasimhan, and Ying Xie (2013), "Treatment effectiveness and side effects: A model of physician learning." Management Science 59, no. 6: 1309–1325.

21. Chintagunta, Pradeep K., Renna Jiang, and Ginger Z. Jin (2009), "Information, learning, and drug diffusion: The case of Cox-2 inhibitors." QME 7, no. 4: 399–443.

22. Zhao, Yi, Sha Yang, Vishal Narayan, and Ying Zhao (2013), "Modeling consumer learning from online product reviews." Marketing Science 32, no. 1: 153–169.

23. Jackson, Matthew O (2008), Social and economic networks. Vol. 3. Princeton, N.J.: Princeton University Press.

24. Mobius, Markus, and Tanya Rosenblat (2014), "Social learning in economics." Annual Reviews of Economics 6, no. 1: 827–847.

25. Goldenberg, Jacob, Barak Libai, and Eitan Muller (2010), "The chilling effects of network externalities." International Journal of Research in Marketing 27, no. 1: 4–15.

26. Tucker, Catherine (2008), "Identifying formal and informal influence in technology adoption with network externalities." Management Science 54, no. 12: 2024–2038.

27. Ryan, Stephen P., and Catherine Tucker (2012), "Heterogeneity and the dynamics of technology adoption." Quantitative Marketing and Economics 10, no. 1: 63–109.

28. Goolsbee, Austan, and Peter Klenow (2002), "Evidence on learning and network externalities in the diffusion of home computers," Journal of Law and Economics, 45, no. 2: 317–343.

29. Iyengar, Raghuram, Asim Ansari, and Sunil Gupta (2007), "A model of consumer learning for service quality and usage," Journal of Marketing Research, 44, no. 4, 529–544.

30. Eagle, Nathan, Alex Sandy Pentland, David Lazer. 2009. Inferring friendship network structure by using mobile phone data. Proceedings of the National Academy of Sciences 106 (36) 15274–15278.

31. Chen, Xinlei, Yuxin Chen, Ping Xiao. 2013. The impact of sampling and network topology on the estimation of social intercorrelations. Journal of Marketing Research 50(1) 95–110.
32. Blondel, Vincent D, Jean-Loup Guillaume, Renaud Lambiotte, Etienne Lefebvre. 2008. Fast unfolding of communities in large networks. Journal of statistical mechanics: theory and experiment 2008(10) P10008.
33. Godinho de Matos, Miguel, Pedro Ferreira, David Krackhardt. 2014. Peer influence in the diffusion of the iphone 3g over a large social network. Management Information Systems Quarterly (Forthcoming).
34. Newman, Mark EJ, Michelle Girvan. 2004. Finding and evaluating community structure in networks. Physical review E 69(2) 026113.
35. Erdem, T. and M. P. Keane (1996), "Decision-making under uncertainty: Capturing dynamic brand choice processes in turbulent consumer goods markets." Marketing Science 15:1–20.
36. Lemon, Katherine, Tiffany White, and Russell Winer (2002), "Dynamic customer relationship management: Incorporating future considerations into the service retention decision," Journal of Marketing, 66(1): 1–14.
37. Pakes, Ariel, Michael Ostrovsky, and Steven Berry (2007), "Simple estimators for the parameters of discrete dynamic games (with entry/exit examples)," RAND Journal of Economics, 38, no. 2: 373–399.
38. Crawford, G. S., and M. Shum (2005), "Uncertainty and learning in pharmaceutical demand." Econometrica 73:1137–1173.
39. Dunne, Timothy, Shawn D. Klimek, Mark J. Roberts, and Daniel Yi Xu (2013), "Entry, exit, and the determinants of market structure." The RAND Journal of Economics 44, no. 3: 462–487.
40. DeGroot, Morris H (2005), Optimal statistical decisions. Vol. 82. John Wiley & Sons.
41. Bajari, Patrick, Lanier Benkard, and Jonathan Levin (2007), "Estimating dynamic models of imperfect competition." Econometrica 75, 1331–1370.
42. Aguirregabiria, Victor, and Pedro Mira (2007), "Sequential estimation of dynamic discrete games." Econometrica 78(2) 1–53.
43. Keane, Michael P., and Kenneth I. Wolpin (1997), "The career decisions of young men." Journal of Political Economy, 105, no. 3: 473–522.

Chapter 11
Network Based Targeting

In this chapter we demonstrate that when an individual's network information is available, we can use the characteristics of network structure to improve the predictive validity when predicting consumer behavior (even in out of sample predictions). However, network structure measures are determined by the network itself, and network is the result of individual's social interactions which is determined by individual characteristics. So we have to develop a structure model of consumer behavior within a network to address the related endogeneity issues.

In 2014, approximately 4 billion people (more than half of the world's population) had used mobile phones. People are spending more time on their mobile phones than ever before. On average, people in the U.S. spent 177 min on their mobile devices every day, compared to 168 min spent on television (Flurry Analytics, comScore). Many people now use mobile phones to connect to the world, and more importantly, to other people.

As the world becomes increasingly dominated by mobile phones, wireless carriers and researchers can make use of the vast amount of mobile usage data to study consumer behavior, social networks, and ultimately the interplay between the two. These datasets contain extensive information at the individual level that is not just about consumer behavior, but also about social networks. However most previous studies using mobile data does not seem to have utilized the full extent of such databases, as they have only focused on information about consumer characteristics and their interactions with the carriers, such as their choice of plans and service use [1–4].

Social networks are essentially the aggregation of social contacts. Within a network, individuals influence each other's behavior through their social contacts. Thus the network structure, which summarizes the nature and pattern of social interaction, can influence user behavior. Figure 11.1 shows the adoption patterns in two networks of the same size in the same city. Using the same network graphing algorithm, one network has the network structure of a tight string ball while the other has a "hub-and-spoke" appearance. Over the same period, 24 individuals (denoted as red dots) adopted the new product in the tight pattern whereas there were only 10 in

© Springer International Publishing AG, part of Springer Nature 2018

Y. Ouyang et al., *Mining Over Air: Wireless Communication Networks Analytics*,

https://doi.org/10.1007/978-3-319-92312-3_11

Fig. 11.1 Network
structure and Samsung Note
II adoption

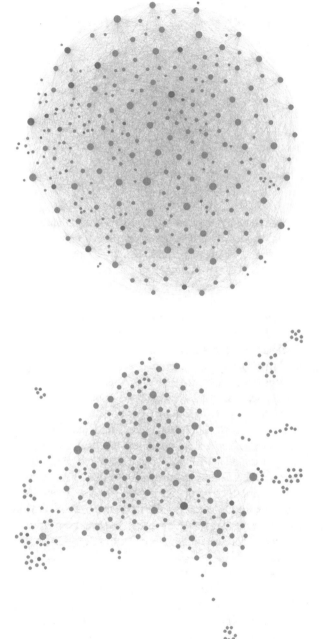

the "hub-and-spoke" pattern. One factor that explains the observed difference in behavior is the network structure. Modeling consumer behavior within the network is also special because of the endogenous issue of individual characteristics on the selection of contacts and ultimately, of network formation.

In this chapter, we propose a modeling approach to study consumer behavior in a network context. The model addresses two challenges that set apart the consideration of behavior within and without a network, which are the effects of the network structure and the endogeneity of the network formation. We will be modeling the individual adoption behavior using the spatial profit model, which was adapted from the classic linear spatial autoregressive (SAR) model [5–8]. We assume that there will be three channels through which the network structure affects individual adoption behavior: (1) the individual characteristics and positions within the network (own and contextual effects); (2) the global network structure as the environment (correlated effect); and (3) peer influence (endogenous effect). For more discussion on these channels of network effects, please refer to Sect. 11.2 of this chapter.

Based on our empirical analysis, we also propose a new targeting strategy, which we call network-based targeting. Traditionally, certain individuals are identified as targets (seeds) based on their individual characteristics. We suggest that, when choosing the targets, it is also important to take into account the network structure information of the focal person's snowball network in a systematic way. To demonstrate the effectiveness of this strategy, data have been collected after 5 months of sampling period to compare with results from conducting out of sample predictions. By comparing the predictions of the product adoption numbers in each person's snowball network (two layers in our case) with the real adoption data, we demonstrate the superior performance of our strategy compared to the common practice in the industry.

Previous studies using mobile data had mainly focused on two separate streams. One stream measures the network structure using the strategy of ego network, which mainly studies how individual position within his local network affects the behavior and product diffusion [9–21]. Individual position reflects how easily one can access and disseminate information through their connections [11, 14]. Examples of standard measures for individual network positions include centrality and the clustering coefficient.

The second stream of research using mobile data investigates the entire network and examines the relationship between global network characteristics and consumer behavior [22–32]. In Economics and Sociology, global network structures had long been used to provide sources to measure a group's social capital (cf., [33–38]). A higher social capital in a group benefits its members economically in many ways, including lower protective transition costs, motivating innovation and accumulation of human capital. Therefore, changes in network structures will affect adoption behaviours, due to changes in accumulation of social capital. Coleman [34] had suggested that the most important source of social capital is network closure, as this circulation of information within a closed network retains and even enhances the quality of information. However, Burt [35] had proposed opposing views, suggesting that "structural holes" or brokerage positions in networks may play an important role in information diffusion and adoption. Thus, in addition to measuring closure, we also assessed the effect of diversity within networks.

In this chapter there are a few questions which we would like to address. First, we would like to introduce a general approach to model consumer behavior within a network. This model integrates both the local and global structures and helps us develop a more comprehensive understanding of how the network structure affects individual behavior. To increase the effectiveness in predicting adoption behavior of clients, we also proposed a network-based targeting strategy. Our proposed model, which we will further discuss in the next sections, provides a way to incorporate both local (based on the ego network) and global (based on the snowball network) network structure measures to model consumer behavior.

Our model consists of two simultaneous equations. In the "network formation" equation, we adopt the correction approach of Hsieh and Lee [8] and correct the network endogeneity problem using latent spatial model; in the "outcome" equation, we use spatial probit model, which is adapted from the classic linear spatial autoregressive (SAR) model. In the SAR model, we incorporate the effects of network structure as individual characteristics, contextual effects, correlated effects, and peer influence.

11.1 Channels of Network Effects

As previously mentioned in the model, we assume that network structures affect individual behaviors through three separate channels. Two of these channels address external influences, and one refers to the individual himself. The first channel refers to individual characteristics and position within the network, including their own effect and contextual effects, such as the user's age and gender. Measures of individual positions in the local network include individual centrality and the individual clustering coefficient, which are based on the ego network perspective. The ego network strategy begins with focal nodes and then tracks down the nodes that are directly connected to the focal ego.

The second channel of network effect is the global network structure as the environment. This network-correlated effect includes measures such as density, global clustering coefficient, eigenvalues, and entropies of the individual centrality. The final channel of interest measures peer influence, also known as the endogenous effect. We define peer influence as an expression of an individual's conformity. To do this, we specify a random coefficient on peer influence as a function of the structure measures from each network community. If the peer influence is positive, an individual is inclined to behave similarly to his/her peers, particularly when many of those peers exhibit the same behavior [39].

The endogeneity issue of network formation is another important factor that is unique to modeling the consumer behavior within a network. When looking at telecommunication data, there are often some unobserved individual characteristics that affect consumer behavior, which may also affect the selection of contacts. These may result in biased estimates of network related variables, such as peer influence and the network structure effects. As previously stated, we define peer influence to be an expression of an individual's conformity. To solve this problem and capture the unobserved individual heterogeneity in the network formation process, we introduce a network formation model following the latent variable approach of Hsieh and Lee [8]. Our model of network formation takes into account the concept of homophily [40], which states that individuals who are similar to one another are more likely to be connected.

In this study, we combine the spatial probit model and network formation model as simultaneous equations. To correct for network endogeneity, we incorporate unobserved latent variables that were obtained from the network formation equation into the outcome equation of the spatial probit model. (The dyad-specific variables that are used in the network formation equation serve as good candidates for natural exclusion restrictions as they do not generally affect the behavioral outcomes, and are valid in many different application contexts.) The proposed model also permits inclusion of ad hoc exclusive restrictions depending on the research context.

When examining a group of individuals with similar behavior in a community, it is especially important to distinguish between peer influence and the homophily effect, where individuals who adopt the same products tend to associate with each other, resulting in network formation. However it may be difficult to separate the effects from each other. We solve this problem by adding the network formation stage to the outcome model. The resulting network formation model is a latent space model, with the center idea that the smaller the distances in observed as well as unobserved characteristics between two people, the more likely they will become friends.

11.2 Dealing with Network Data

In the empirical application, the model was applied to a proprietary dataset that we collected from one of the major Greater Asia mobile carriers, in two medium-sized cities in China. The time period of data collection started from October 2012 and went on till May 2013, which corresponded with the release of the Samsung Note II. We used a two-layer snowball sampling with the adopters of Samsung Note II during the sample period. Each adopter acts as a focal seed to extract the relevant data from the entire customer base of the cities. As there is the possibility that the networks from snowball sampling will overlap, we further use the Louvain method of community construction [41] to extract non-overlapping networks from the snowball sample. This model is favorable for its efficiency compared to other existing methods and may be ideal for analyzing big datasets from telecommunications companies. To control for product level heterogeneity and to reflect the product hierarchy, we consider individual adoptions of Samsung cellphones at three choice levels: (1) the newest mobile phone model Samsung Note II, (2) Samsung high brand-tier (high-end) phones, and (3) brand level (all Samsung brand phones). This three-level hierarchical structure allows us to observe how the estimates vary with respect to the product definitions.

Our preliminary empirical analysis had provided us with several results that are worth-noting. Firstly, both local and global network structure measures have significant effects on the phone adoption behavior within a network. When the network structure effects are not properly controlled, omitted variable biases occur in the estimates of the other variables. Second, each structure measure exhibits different effects on different product definitions. Third, the similarities in the demographics and unobserved factors significantly affect the formation of networks.

However, even after correcting for network endogeneity, the estimates of network effects only exhibit mild differences as compared to before. This suggests that the correlation between network formation and adoption behavior through unobservable is not strong. Lastly, among all of the measures, the global clustering coefficient exhibits the strongest effect on behavior, and thus is selected if we were asked to suggest one global network measure.

Based on these results, we have proposed an alternative targeting strategy called network-based targeting. To demonstrate the effectiveness of our strategy, we collect new data after 5 months of sampling period and conduct out of sample predictions. Following, we will demonstrate the superior performance of this strategy compared to the common practice in the industry. The difference in predictive validity is particularly pronounced in the case of the Samsung Note II where the adoption is very sparse in the data.

11.3   Modeling Strategies Issues

We propose an empirical approach for modeling the consumer behavior within a network. This model intends to study relationship between dynamic network structures and peer influence. We begin by introducing the linear spatial autoregressive (SAR) model which is our initial model. This model is adapted to the binary choice decisions, the resulting model being called spatial profit model. It captures the effects of an individual's own characteristics and the contextual effect from the peers' characteristics as well as the endogenous peer effect.

Then, the model is extended to take into account local and global network effects. For individual local network measures, we consider three centrality measures: degree, betweenness, and eigenvector, and the individual clustering coefficient. For global network structure measures, we include network size, density, global clustering coefficient, assortativity, maximum and minimum eigenvalues, and three entropy measures based on degree, betweenness, and eigenvector centralities. This list of measures is not meant to be comprehensive but target ones that have been mentioned or studied separately in different fields (e.g., sociology, economics, and epidemiology) but have not been examined together to provide empirical support. At this point, the reader can recall Sect. 9.3 describing network measures.

Finally, the model needs to handle endogeneity of network formation, that is to "fix" the homophily effect of the classic model. This correction approach follows [8]. In a nutshell, the model is adjusted using latent spatial model, by adding random coefficient specifications on the endogenous effect in terms of network structures.

11.3.1 Linear Spatial Autoregressive Mode

We assume the different networks have been already constructed for the population, based on social mobile phone interconnections. In other words, individuals belong to pre-specified groups in our environment. Let $g = \{1, \ldots, G\}$ denote the index of groups, and the size of each group g ism_g. Individuals are indexed with i for each interconnected group, and each individual belongs only to one group. We denote the behavioral outcome by $y_{i,g}^*$ and $Y_g^* = \left(y_{1,g}^*, \ldots, y_{m_g,g}^* \right)'$ is a collection of m_g outcome variables in groupg. The network of group g is represented by a $m_g \times m_g$ spatial weight (network adjacency) matrixW_g, where each (i, j) element $w_{ij;\,g}$ equals one if customer i links to customer j and zero otherwise. We set $w_{ii;\,g} = 0$ for all i and the matrix W_g is assumed to be symmetric, i.e., $w_{ij;\,g} = w_{ji;\,g}$.

We use $x_{i;\,g}$ to denote customer exogenous characteristics, which is a vector of R^P. This vector contains variables such as the gender, the age or the current smartphone of the individual. We also define$X_g = \left(x_{1,g}, \ldots, x_{m_g,g} \right)'$.

We begin by introducing the linear SAR model and its adaptation to the binary choice decisions, and then discuss our extensions that allow for the various afore-mentioned network effects. A standard SAR model for $y_{i,g}^*$ is given by:

$$y_{i,g}^* = \lambda \sum_{\substack{j \neq i}}^{m_g} \frac{w_{ij;g}}{\sum_{j \neq i} w_{ij;g}} y_{j,g}^* + x_{i,g}\beta_1 + \sum_{\substack{j \neq i}}^{m_g} \frac{w_{ij;g}}{\sum_{j \neq i} w_{ij;g}} x_{j,g}\beta_2 + \alpha_g + \varepsilon_{i;g} \qquad (11.1)$$

In Eq. (11.1), coefficient λ captures the endogenous (peer) effect. The existence of a positive endogenous effect can trigger a multiplier effect that facilitates the diffusion of behavior throughout the network links. Coefficients β_1 and β_2 capture the effects of an individual's own characteristics and the contextual effect from the peers' characteristics, respectively. In the conventional social interaction model, i.e., the linear-in-means model, individuals by default interact with all of their peers (except themselves) in the same group. This imposes a homogeneous peer composition across individuals. Due to the lack of individual variation, the linear-in-means model may not differentiate between the endogenous and contextual effects, which are known as the "reflection problem" [42]. When incorporating network information into the social interaction model through the SAR specification, Bramoullé et al. [6] show that the endogenous and contextual effects can be separately identified in Eq. (11.1), as long as the peers of individuals do not perfectly overlap.

In addition to the refection problem, the existence of correlated effects from the group unobservables also causes an identification problem in the social interaction model. An example of this problem can be observed from studying the peer influences among school students. If the empirical model fails to control school environmental factors, such as the classroom facilities and teacher qualities, the estimate of the endogenous effect will be biased due to the correlation with the error term that absorbs these school correlated effects [43–45]. To handle this problem, we specify a group-specific effect α_g in Eq. (11.1), which captures the unobserved group characteristics in group g. The individual error term $\varepsilon_{i;\ g}$ should be uncorrelated with the regressors, following a normal distribution with a zero mean and varianceσ_ε^2.

In the case where the behavior variable is a zero-one binary choice, such as the product adoption denoted by $y_{i,\ g}$, we can model the utility of adoption by $y_{i,g}^*$ of Eq. (11.1). By specifying the following relationship between $y_{i,g}^*$ and the observed behavior variable$y_{i,\ g}$,

$$y_{i,g} = \begin{cases} 1 & \text{if } y_{i,g}^* > 0 \\ 0 & \text{otherwise} \end{cases} \tag{11.2}$$

we can transform the linear SAR model into a spatial probit model. In terms of estimation, we normalize σ_ε^2 as one under the spatial probit model.

By defining $\widehat{W_g}$ as the row-normalized W_g, $\varepsilon_g = \left(\varepsilon_{1,g}, \ldots, \varepsilon_{m_g,g}\right)'$ and 1 as the column vector of 1 of size m_g, Y_g^* can be rewritten under the following matrix form:

$$Y_g^* = \lambda \widehat{W_g} Y_g^* + X_g \beta_1 + \widehat{W_g} X_g \beta_2 + \alpha_g 1 + \varepsilon_g \tag{11.3}$$

11.3.2 Network Interaction Model

To model the consumer behavior within a network, we extend the SAR model in Eq. (11.3) by incorporating local (from ego's perspective) and global network effects in the following way. First, for the local effects, we include local network measures such as the individual centrality measures and clustering coefficient into the specification of individual's own characteristics $x_{i,\ g}$ and those of his/her peers'. Second, for the global effects, we specify a L-dimensional vector of global network characteristics$S_g = (S_{1,\ g}, \ldots, S_{L,\ g})'$, with each $S_{1,\ g}$ denoting measures such as the density, global clustering coefficient, and assortativity, and then replace α_g in Eq. (11.1) by $S_g \delta + v_g$. δ is a coefficient in R^k where k in the number of global network effect characteristics. By construction, it reflects the correlated effects from the different global network measures. The term v_g is a new group-specific error, which is a random effect with variance σ_v^2 that is uncorrelated with all of the other regressors in the model. This specification is similar to the correlated random effect in Mundlak [46] and Chamberlain [47].

Finally, we could get the network interaction model is (for $g \in \{1, \ldots, G\}$, where $\widehat{W}_g$ denotes the row-normalized W_g and 1 denotes a $m_g \times 1$ vector of ones):

$$Y_g^* = \lambda \widehat{W}_g Y_g^* + X_g \beta_1 + \widehat{W}_g X_g \beta_2 + \left(S_g \delta + v_g\right)1 + \varepsilon_g \qquad (9.4 = \text{Case III})$$

$$y_{i,g} = 1_{y_{i,g}^* > 0}$$

11.3.3 Endogenous Peer Effects

The other extension provided by our study is to introduce group heterogeneity into the endogenous peer effect. This extension considers that the global network structures not only affect individual adoptions through the group correlated effects, but also moderate the endogenous peer effect in each network. We reflect this feature in our model by specifying a random coefficient on the endogenous effect and changing a fixed λ into $\lambda_g = \lambda_0 + S_g \kappa + u_g$, where κ denotes a $L \times 1$ vector of coefficients and u_g is a random normal disturbance with a mean zero and variance σ_u^2. Given the above extensions, the final model that we use to capture the network effects on consumer behavior can be presented in the following vector-matrix form (for $g \in \{1, \ldots, G\}$):

$$Y_g^* = \left(\lambda_0 + S_g \kappa + u_g\right)\widehat{W}_g Y_g^* + \alpha_0 1 + X_g \beta_1 + \widehat{W}_g X_g \beta_2 + \left(S_g \delta + v_g\right)1 + \varepsilon_g$$

$$(11.5)$$

The endogeneity of the network matrix W_g is a challenge in modeling the behavior in a network, as it prevents us from obtaining unbiased coefficient estimates from Eq. (11.5). Unlike the unobserved group heterogeneity, which is addressed by the group correlated effect, this problem arises from the unobserved individual heterogeneity that affects both the adoption behavior and friendship selection. For example, an individual's attitude toward freshness and excitement can affect his/her choice of friends and behavior. When these relevant but unobserved characteristics are omitted from the model, the repressors that are constructed from the network will potentially correlate with the error term ε_g and become endogenous. To correct this omitting variable bias, we follow Goldsmith-Pinkham and Imbens [48] and Hsieh and Lee [8] by introducing individual latent variables $z_{i,g}$ to control the unobserved characteristics in both the network link formation and outcome interaction models. This approach asserts that as long as the unobserved individual characteristics are controlled in the model, the problem of network endogeneity can be resolved.

At the center of the modeling network link formation, we rely on the concept of "homophily" [40], which states that individuals with similar characteristics are more likely to become friends. To reflect the homophily in the model, we consider regressors from the observed dyad-specific characteristics and the distances of the latent variables between individuals. The dyadspecific characteristics are captured by either dummy variables or continuous variables that represent the similarity (or difference) between two individuals with respect to the specific characteristics. For example, if two individuals are of the same gender, then the dyad gender dummy will be one, and zero otherwise. Yang and Allenby [49] used the same approach to model the dependence of consumer preference based on observed characteristics. The distance of the latent variables captures the homophily of the unobserved characteristics. The smaller the difference between individuals' unobserved characteristics, the more likely they will become friends. Accordingly, we model the link formation by the following logit probability specification:

$$P\left(w_{ij,g} | c_{ij,g}, z_{i,g}, z_{j,g}\right) = \left(\frac{\exp\left(\psi_{ij,g}\right)}{1 + \exp\left(\psi_{ij,g}\right)}\right)^{w_{ij,g}} \left(\frac{1}{1 + \exp\left(\psi_{ij,g}\right)}\right)^{1-w_{ij,g}} \quad (11.6)$$

$$\psi_{ij,g} = c_{ij,g}\gamma + \eta_1 \left|z_{i1,g} - z_{j1,g}\right| + \ldots + + \eta_{\bar{d}} \left|z_{i\bar{d},g} - z_{j\bar{d},g}\right|.$$

In Eq. (11.6), variable $c_{ij,g}$ is a R-dimensional vector of observed dyad-specific characteristics; Individual unobserved (latent) variables $z_{i,g}$ are assumed to be multidimensional, and $|z_{id,g} - z_{jd,g}|$ captures the distance on the d-th unobserved characteristic between individuals (i, j). As a result, coefficients $\{\eta_d\}_{d=1\bar{d}}$ are expected to be negative. The model of Eq. (11.6) is a variation of the "latent Space" model of network formation from Statistics [50].

Linking Eqs. (11.6) to (11.5), we introduce latent variables $Z_g = \left(z_{1,g}, \ldots, z_{m_g,g}\right)'$ to Eq. (11.5) to control for the individual unobservable. We finally obtain the selection-corrected SAR model (for $g \in \{1, \ldots, G\}$):

$$Y_g^* = \left(\lambda_0 + S_g\kappa + u_g\right)\widehat{W_g}\,Y_g^* + l_g\alpha_0 + X_g\beta_1 + \widehat{W_g}\,X_g\beta_2 + Z_g\rho$$
$$+ l_g\left(S_g\delta + v_g\right) + \xi_g \quad (11.7)$$

where ξ_g follow i.i.d. $N\left(0, Id_{m_g}\right)$. The way Eq. (11.7) corrects for endogeneity bias is in line with the control function approach (see survey in [51]), and the selection model in Heckman [52] in some respects. The validity of the Heckman-type selection correction method relies on the error term assumptions in the outcome and selection equations, and the exclusion restriction on the exogenous variables.

In our link formation and outcome SAR system, we assume that the conditional expectation of $\varepsilon_{i,g}$ given $z_{i,g}$ is linear, and $z_{i,g}$ is from a normal distribution [53]. Therefore, ε_g in Eq. (11.5) can be decomposed into $Z_g\rho + \xi_g$ in Eq. (11.7). Since there is no closed form expression of Z_g on exogenous variables available from the link formation equation, we cannot implement a two-stage (limited information likelihood) estimation by first estimate Z_g in the first step and then plug it into Eq. (11.7) in the second step. Alternatively, we estimate Eqs. (11.6) and (11.7) based on the full information likelihood approach.

Theoretically, given the parametric error term assumptions, we can identify the model parameters without the exclusion restrictions on the exogenous variables. However, the exclusion restrictions are essential for guaranteeing identification of the model when error term assumptions are removed. The model proposed in this paper allows the inclusion of ad hoc exclusion restrictions. However, natural exclusion restrictions are embedded in our model system, which is, the exogenous dyad specific variables, i.e., $c_{ij,\,g}$, used in the link formation model. The dyad variables are naturally excluded from the outcome interaction model, where only individual-specific variables enter. Thus, even in a research context where ad hoc nonparametric exclusion conditions are difficult to find, the proposed approach still indicates a way to use parametric conditions to alleviate the identification problem.

11.4 Findings and Applications

11.4.1 Explanation of Results

It is worth mentioning that although the current link formation model explains the unobserved individual heterogeneity during the formation process, two features are still missing from the model. The first feature is the homophily of behavior where individuals tend to associate with those who behave similarly. The current model is designed to analyze cross-sectional behavior and network outcomes, and therefore has difficulty distinguishing peer influence from the homophily of behavior [54]. An alternative model proposed by Snijders et al. [55] uses panel information on networks and behavior to differentiate peer influence and homophily. However, in their model, the homophily of the unobservables becomes uncontrolled. The second missed feature is the dependence among links. An obvious example of link dependence is the transitivity of friendships. Specifically, when individuals i and j are both friends of individual k, i and j are likely to be friends too. Allowing the link dependence abruptly complicates the link formation process because the conditional probability in Eq. (11.6) should be changed to $P(w_{ij,\,g} | c_{ij,\,g}, z_{i,\,g}, z_{j,\,g}, W_{-ij,\,g})$, with $W_{-ij,\,g}$ denoting all links in W_g except $w_{ij,\,g}$. However, due to the coherency issue, we usually cannot compute the joint probability of W_g directly from $\prod_{i,j} P(w_{ij,g} | c_{ij,g}, z_{i,g}, z_{j,g}, W_{-ij,g})$ and thus the estimation becomes difficult. In the literature, the standard model for explaining link dependence is the exponential random graph model (ERGM) [56]. However, ERGMs typically explain link dependence in a mechanical way without considering the strategic individual decisions. Incorporating these two missing features into the current link formation model of Eq. (11.6) is an important extension that we will explore in our future work.

11.4.2   Network Based Targeting

Correctly targeting customers and customizing to their needs has always been a general focus in marketing. Companies have long realized the value of providing the right product or service to the right customer. To do this, companies have developed an index of its customers based on some prediction models using their individual characteristics, such as their age, gender, tenure, location etc. Customers are generally ranked accordingly, and customers with higher rankings will be more likely to be a target of receiving promotional or commercial messages.

However, the social networks of customer can also provide many important clues to predicting their behaviors. Our study demonstrated how the wireless carrier can use social network information to better predict customer behavior and to initiate a more effective targeting campaign. In order to demonstrate the effectiveness of our strategy, we collected a new sample between Octobers to December 2013, 5 months after our original sample. Our goal is to compare the out-of-sample predictions, with different methods, with the customers' true adoption behavior, and to provide an applicable tool for companies to implement more effective targeting strategies.

Our analysis had extracted 6227 customers, who have overlap with the original sample collected from one of the cities. Using these customers, wireless carries can construct 6227 two-layer snowball networks, and customers are ranked according to the predicted number of adoptions in their network. We then explore the three separate approaches to predict adoption behaviors. The first is based on the profit model used in Case (1). This represents the currently most abundant prediction technique used in the business world. The second approach is also based on the profit model, however, there are extended controls on the local and global network structures measures, as reported in Case II. The third approach is based on our proposed model, the random coefficient spatial profit model. In this prediction exercise we do not include an intensive control on the global network measures as we did in estimation so the excersice is closer to what the company will do in reality. Instead, we only use the global clustering coefficient, as it is found to be the most important adoption facilitator.

We apply each model with the coefficients found in the estimation to the new sample of customers and their networks, and predict the values of $y_{i,g}^*$ and the predicted adoptions according to Eq. (11.2). The customers are ranked based on the scores calculated by dividing the number of predicted adoptions in an individual's network by the highest number of predicted adoptions found in all networks (thus bounded between zero and one).

We plot the real numbers of adoptions in the new sample against the customer scores calculated from the three models in Fig. 11.2. The three panels from top to bottom represent the case of the Samsung Note II, Samsung high-end, and Samsung brand, respectively. Within each panel, the three columns from left to right represent the probit, probit with network information, and spatial probit model approaches, respectively. If the predictions are accurate, i.e., the ranking of a customer reflects

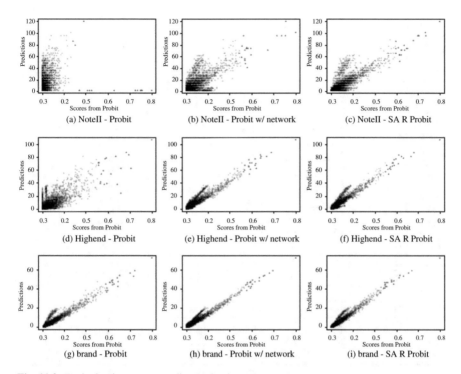

Fig. 11.2 Real adoption versus predicted adoption

the real number of adoptions, the scatter plots should fall on a positive-sloped (45 degree if properly scaling two axes) straight line. The dots highlighted in red represent the top 20 ranked individual based on each prediction method.

We compared the performance of the three different approaches in two ways. First we compared the predictive validity between models, and found the most signifiant difference in performance in adoption of the Samsung Note II, as shown in the first panel. When the real adoption rate is low and the data variation is sparse, the probit model can barely make the prediction for 5 months later, with few dots falling on the 45 degree line. In contrast, the probit model with network information and the spatial probit model work properly and the dots roughly fall on the 45 degree straight line. This result shows that adding network information to the prediction can significantly improve the prediction accuracy. The performance of the profit model improves when the adoption rates increases and data variation is no longer sparse. However, the extended probit model still has superior performance with network controls, showing that network information is useful in reducing prediction error. Finally, the proposed spatial probit model further reduces the prediction error in all three panels and performs even better than the probit model with network information.

Table 11.1 Network based targeting – seeds selection

NoteII			Highend			Brand		
Probit	Probit w/net.	S-Probit	Probit	Probit w/net.	S-Probit	Probit	Probit w/net.	S-Probit
4134	2	1	1	1	1	1	1	1
3765	1	3	17	2	2	2	2	2
3965	4	2	3	4	4	5	3	3
4066	3	4	2	5	6	3	4	5
3764	5	5	18	6	5	4	5	4
4133	22	6	300	3	3	10	6	6
3766	33	8	93	17	12	17	8	8
3394	17	15	4	12	7	6	10	15
3859	16	16	32	7	8	15	7	9
4135	172	20	86	13	13	7	15	10
4545	13	10	6	10	11	11	17	7
4415	15	21	7	21	10	16	9	12
3977	18	22	97	11	15	8	21	18
3914	109	31	24	31	21	13	12	11
4069	21	81	34	8	16	12	13	17
1	49	32	75	16	17	28	18	14
4008	62	33	84	15	31	21	11	16
4076	6	13	9	14	20	31	14	21
10	11	17	14	23	14	25	28	13
9	42	24	11	26	26	14	16	22

Second, to illustrate how the prediction can affect the effectiveness of targeting, we show the target lists of the top 20 ranked individuals according to the three approaches in Table 11.1. The ID number assigned to each individual network is based on the ranking of the real data. The smaller the number, the higher the rank which meant the more real adoptions we observe within the individuals network. Therefore, the target list should comprise only the numbers from 1 to 20. The more numbers between that range that are included in the list, the more effective the targeting strategy is. Consistent with Fig. 11.2, in the case of the Samsung Note II, the probit model's target list misses as many as 17 out of the 20 targets. Moreover, the probit model with network information and the proposed spatial probit model only miss 5–6 targets. As the adoption rates increase, the target list of the probit model improves, as does that of the other two approaches.

In this study, we demonstrated that targeting strategies that rely on accurate out-of-sample predictions can be quite sensitive when they are based on reduced form models that only use individual characteristics. For practitioners, it may not be feasible to use such a sophisticated model as we propose here. However, simply adding the network information can significantly improve the predictive validity as well as the effectiveness of the targeting strategy.

References

1. Iyengar, Raghuram, Asim Ansari, Sunil Gupta. 2007. A model of consumer learning for service quality and usage. Journal of Marketing Research 44(4) 529–544.
2. Grubb, Michael D. 2012. Dynamic nonlinear pricing: biased expectations, inattention, and bill shock. International Journal of Industrial Organization 30(3) 287–290.
3. Ascarza, Eva, Anja Lambrecht, Naufel Vilcassim. 2012. When talk is free: The effect of tariff structure on usage under two-and three-part tariffs. Journal of Marketing Research 49 (6) 882–899.
4. Gopalakrishnan, Arun, Raghuram Iyengar, Robert J Meyer. 2014. Consumer dynamic usage allocation and learning under multipart tariffs. Marketing Science 34(1) 116–133.
5. Lee, Lung-Fei. 2007. Identification and estimation of econometric models with group interactions, contextual factors and fixed effects. Journal of Econometrics 140(2) 333–374.
6. Bramoullé, Yann, Habiba Djebbari, Bernard Fortin. 2009. Identification of peer effects through social networks. Journal of econometrics 150(1) 41–55.
7. Lin, Xu. 2010. Identifying peer effects in student academic achievement by spatial autoregressive models with group unobservables. Journal of Labor Economics 28(4) 825–860.
8. Hsieh, Chih-Sheng, Lung-Fei Lee. 2016. A social interactions model with endogenous friendship formation and selectivity. Journal of Applied Econometrics 31(2) 301–319.
9. Hill, Shawndra, Foster Provost, Chris Volinsky. 2006. Network-based marketing: Identifying likely adopters via consumer networks. Statistical Science 256–276.
10. Tucker, Catherine. 2008. Identifying formal and informal influence in technology adoption with network externalities. Management Science 54(12) 2024–2038.
11. Goldenberg, Jacob, Sangman Han, Donald R Lehmann, Jae Weon Hong. 2009. The role of hubs in the adoption process. Journal of Marketing 73(2) 1–13.
12. Nair, Harikesh S, Puneet Manchanda, Tulikaa Bhatia. 2010. Asymmetric social interactions in physician prescription behavior: The role of opinion leaders. Journal of Marketing Research 47 (5) 883–895.
13. Trusov, Michael, Anand V Bodapati, Randolph E Bucklin. 2010. Determining influential users in internet social networks. Journal of Marketing Research 47(4) 643–658.
14. Stephen, Andrew T, Olivier Toubia. 2010. Deriving value from social commerce networks. Journal of marketing research 47(2) 215–228.
15. Hinz, Oliver, Bernd Skiera, Christian Barrot, Jan U Becker. 2011. Seeding strategies for viral marketing: An empirical comparison. Journal of Marketing 75(6) 55–71.
16. Katona, Zsolt, Peter Pal Zubcsek, Miklos Sarvary. 2011. Network effects and personal influences: The diffusion of an online social network. Journal of Marketing Research 48(3) 425–443.
17. Iyengar, Raghuram, Christophe Van den Bulte, Thomas W Valente. 2011. Opinion leadership and social contagion in new product diffusion. Marketing Science 30(2) 195–212.
18. Narayan, Vishal, Vithala R Rao, Carolyne Saunders. 2011. How peer influence affects attribute preferences: A bayesian updating mechanism. Marketing Science 30(2) 368–384.
19. Nitzan, Irit, Barak Libai. 2011. Social effects on customer retention. Journal of Marketing 75 (6) 24–38.
20. Ugander, Johan, Lars Backstrom, Cameron Marlow, Jon Kleinberg. 2012. Structural diversity in social contagion. Proceedings of the National Academy of Sciences 109(16) 5962–5966.
21. Yoganarasimhan, Hema. 2012. Impact of social network structure on content propagation: A study using youtube data. Quantitative Marketing and Economics 10(1) 111–150.
22. Jackson, Matthew O., Leeat Yariv. 1996. Social networks and the diffusion of economic behavior. Yale Economic Review 3(2) 42–47.
23. Jackson, Matthew O, Brian W Rogers. 2007. Relating network structure to diffusion properties through stochastic dominance. The BE Journal of Theoretical Economics 7(1).

24. Bampo, Mauro, Michael T Ewing, Dineli R Mather, David Stewart, Mark Wallace. 2008. The effects of the social structure of digital networks on viral marketing performance. Information Systems Research 19(3) 273–290.
25. Centola, Damon. 2010. The spread of behavior in an online social network experiment. Science 329(5996) 1194–1197.
26. Eagle, Nathan, Michael Macy, Rob Claxton. 2010. Network diversity and economic development. Science 328(5981) 1029–1031.
27. Dover, Yaniv, Jacob Goldenberg, Daniel Shapira. 2012. Network traces on penetration: Uncovering degree distribution from adoption data. Marketing Science 31(4) 689–712.
28. Libai, Barak, Eitan Muller, Renana Peres. 2013. Decomposing the value of word-of-mouth seeding programs: Acceleration versus expansion. Journal of marketing research 50 (2) 161–176.
29. Schlereth, Christian, Christian Barrot, Bernd Skiera, Carsten Takac. 2013. Optimal product sampling strategies in social networks: How many and whom to target? International Journal of Electronic Commerce 18(1) 45–72.
30. Peres, Renana. 2014. The impact of network characteristics on the diffusion of innovations. Physica A: Statistical Mechanics and its Applications 402 330–343.
31. Aral, Sinan, Dylan Walker. 2014. Tie strength, embeddedness, and social influence: A large-scale networked experiment. Management Science 60(6) 1352–1370.
32. Bramoullé, Yann, Rachel Kranton, Martin D'Amours. 2014. Strategic interaction and networks. The American Economic Review 104(3) 898–930.
33. Granovetter, Mark. 1985. Economic action and social structure: the problem of embeddedness. American journal of sociology 481–510.
34. Coleman, James S. 1988. Social capital in the creation of human capital. American journal of sociology S95–S120.
35. Burt, Ronald S. 1993. The social structure of competition. Explorations in economic sociology 65: 103.
36. Nahapiet, Janine, Sumantra Ghoshal. 1998. Social capital, intellectual capital, and the organizational advantage. Academy of management review 23(2) 242–266.
37. Glaeser, Edward L, David I Laibson, Jose A Scheinkman, Christine L Soutter. 2000. Measuring trust. Quarterly Journal of Economics 811–846.
38. Karlan, Dean, Markus Mobius, Tanya Rosenblat, Adam Szeidl. 2009. Trust and social collateral. The Quarterly Journal of Economics 124(3) 1307–1361.
39. Young, H Peyton. 2009. Innovation diffusion in heterogeneous populations: Contagion, social influence, and social learning. The American economic review 99(5) 1899–1924.
40. Lazarsfeld, Paul F, Robert K Merton, et al. 1954. Friendship as a social process: A substantive and methodological analysis. Freedom and control in modern society 18(1) 18–66.
41. Blondel, Vincent D, Jean-Loup Guillaume, Renaud Lambiotte, Etienne Lefebvre. 2008. Fast unfolding of communities in large networks. Journal of statistical mechanics: theory and experiment 2008(10) P10008.
42. Manski, Charles F. 1993. Identification of endogenous social effects: The reflection problem. The review of economic studies 60(3) 531–542.
43. Hoxby, Caroline. 2000. Peer effects in the classroom: Learning from gender and race variation. Tech. rep., National Bureau of Economic Research.
44. Hanushek, Eric A, John F Kain, Jacob M Markman, Steven G Rivkin. 2003. Does peer ability affect student achievement? Journal of applied econometrics 18(5) 527–544.
45. Fletcher, Jason M. 2010. Social interactions and smoking: Evidence using multiple student cohorts, instrumental variables, and school fixed effects. Health Economics 19(4) 466–484.
46. Mundlak, Yair. 1978. On the pooling of time series and cross section data. Econometrica: journal of the Econometric Society 69–85.
47. Chamberlain, Gary. 1982. Multivariate regression models for panel data. Journal of Econometrics 18(1) 5–46.

48. Goldsmith-Pinkham, Paul, Guido W Imbens. 2013. Social networks and the identification of peer effects. Journal of Business & Economic Statistics 31(3) 253–264.
49. Yang, Sha, Greg M Allenby. 2003. Modeling interdependent consumer preferences. Journal of Marketing Research 40(3) 282–294.
50. Hoff, Peter D, Adrian E Raftery, Mark S Handcock. 2002. Latent space approaches to social network analysis. Journal of the american Statistical association 97(460) 1090–1098.
51. Navarro, Salvador. 2008. Control functions. The new Palgrave dictionary of economics.
52. Heckman, J. J. 1979. Sample selection bias as a specification error. Econometrica: Journal of the econometric society 153–161.
53. Olsen, Randall J. 1980. A least squares correction for selectivity bias. Econometrica: Journal of the Econometric Society 1815–1820.
54. Hartmann, Wesley R, Puneet Manchanda, Harikesh Nair, Matthew Bothner, Peter Dodds, David Godes, Kartik Hosanagar, Catherine Tucker. 2008. Modeling social interactions: Identification, empirical methods and policy implications. Marketing letters 19(3–4) 287–304.
55. Snijders, Tom, Christian Steglich, Michael Schweinberger. 2007. Modeling the coevolution of networks and behavior. Han Oud Kees van Montfort, Albert Satorra, eds., Longitudinal models in the behavioral and related sciences. Lawrence Erlbaum, 41–71.
56. Robins, Garry, Tom Snijders, Peng Wang, Mark Handcock, Philippa Pattison. 2007. Recent developments in exponential random graph (p*) models for social networks. Social networks 29 (2) 192–215.

Chapter 12
Social Influence and Dynamic Network Structure

In the previous chapter we discussed how social influence can affect product adoption. In this chapter, we want to further investigate the interplay between network structures and effect of social influence to facilitate diffusion. The role social influence plays in diffusion of new products is often studied for its multiplier effects, which can be helpful in facilitating the diffusion process [1]. These are especially of interest to companies, who can take advantage of social influence in developing marketing strategies.

Individual behavior involves interaction with other individuals, and eventually all of these interactions make up entire social networks. However, outcome of interactions largely depends on the entire network structure. Watt and Dodds [2] had found that the ability of any individual to trigger a cascade depends more on the global network structure than on the personal degree of that individuals' influence. For example, a community with diverse ties is correlated with better economic development than a community with homogenous ties [3].

In this study, we again focus on the particular form of interaction called social influence. Social influence expresses an individual's conformity motive in adopting a behavior similar to that of his/her peers, particularly when that behavior is exhibited by many of those peers [4]. As previously suggested, the adoption behavior of an individual does not simply change upon exposure to other adopters, instead it is assumed to be rationalized. Theoretically, well-connected individuals have a high propensity to be influenced by friends. Densely-connected networks consequently will have high changes of exerting social influence, resulting in strong behaviour diffusion between network members.

Social influence should therefore be able to explain why network positions (characteristics) matter in determining the individual outcomes. Previous studies in the past had mostly used measures of network structures to explain the outcomes of individual adoption decisions. Here, our main focus is on the process of product diffusion. The understanding of the mechanism behind correlation of network structure and adoption behavior will enable researchers and companies to

Y. Ouyang et al., *Mining Over Air: Wireless Communication Networks Analytics*,
https://doi.org/10.1007/978-3-319-92312-3_12

developing more efficient marketing strategies and take better advantage of the previous findings.

For data collection, we perform snowball sampling based on the Call Detail Records (CDRs) of users. This method was demonstrated to perform well in recovering social inter-correlations, as it preserves the network structure [8]. Next, we identify networks that exhibit social influence by using a stochastic actor-based dynamic network model. The challenge here, again, is to distinguish social influence, where peer adoption decisions affect the decision of the individual, from homophily, where individuals who adopt the same products tend to associate with each other.

Aral et al. [9] proposed using dynamic network information and propensity matching to solve this problem of separating social influence from homophily. However, this is not the best solution in this study as it considers the entire population as one single network, which renders comparisons of network structure infeasible. It is also restrictive, due to the large number of samples within a network that is required for effective matching in this approach. Another instrumental variable approach that is typically used also suffers from a similar limitation in addressing the endogeneity problem, in that the effectiveness of the instrumental variable relies on variations in the entire network.

The stochastic actor-based dynamic network model solves this issue [10–12]. This model studies the co-evolution of network formation and adoption behaviours. Using longitudinal network information we were able to identify and quantify the social influence effect.

Finally, meta-regression analysis was used to identify factors that shape social influence by comparing structural differences between networks. Using meta-regression analysis, we aim to link network structure characteristics to the existence and magnitude of social influence within the network. In studying social influence, we propose the following list of network structure measures, listed in Table 12.1. We focus on measures that have been previously discussed either empirically [2, 5–7], or theoretically [13, 14]. Some of these measures are micro-level measures based on node positions (e.g., centrality of initial adopters), whereas others are macro-level measures that provide a birds-eye view of the entire network.

Data for the study was collected from a mobile carrier in China, with approximately 1.36 million users in their customer base. The data collection period corresponded with the release of the Samsung Note II, which began from October 2012 to May 2013, and spanned two medium sized cities in China. Such data has shown to be able to accurately predict cognitive constructs such as friendship [15]. Similar to the previous chapter, we consider three choice levels to develop a structural perspective on network influence relationship: adoption of the new Samsung Note II, adoption at high brand-tier level and lastly at the brand level.

For our sampling strategy, we first identify the Samsung Note II adopters from November 2012 to May 2013 as the focal point of each social network. Then we conduct a two-layer snowball-sampling, with the customers who have been in contact with the focal point within these last 7 months in composing the first layer, and customers who have been in contact with customers in the first layer comprise

Table 12.1 Network measure

Network measure (used to operationalize question)	Question answered
Network size and density	Large network vs. enhanced network connections, which are more likely to exhibit social influence
Network structure entropy	Between an evenly distributed network and a network dominated by some nodes with high centrality, which one will promote diffusion
Standard deviation of the edge numbers across time	Whether network dynamics affect social influence
Clustering coefficients	If the friends of an individual are also friends, do they therefore have a larger influence on the individual
Minimum eigenvalue of adjacency matrix	Is social influence somehow related to the equilibrium notion in social networking games
Epidemic threshold and assortativity	Should networks with social influence be targeted for viral marketing
Ratio of inward versus outward network edges	How do asymmetrically connected local clusters affect social influence
Degree centrality and eigenvector centrality of initial adopters	Do differently located initial adopters trigger different processes

the second layer. In the end, we obtained a sample of 26,000 customers from 1083 social networks with a mean network size of 110. We construct seven time-varying monthly network matrices from November 2012 to May 2013 based on the monthly calling and SMS record of each individual within the network. In each matrix, an edge is placed between two individuals to determine whether they have called or texted each other within the same month. Thus, network edges are undirected, and the resulting matrix is symmetric. We measure network statistics based on the accumulated network, which includes all edges across sampling periods.

In our data, Samsung has 83 models with the carrier; 24 of these models are labeled as "high-end" models by the carrier. The average price of Samsung models is 2522 RMB, with actual prices ranging from 199 RMB to 10,600 RMB; the average price of the high-end models is 4549 RMB. At the time of its release, Samsung Note II was sold for 5699 RMB, while iPhone 4S was sold for 4488 RMB. Figure 12.1 shows that eight people adopted the Samsung Note II in November 2012 right after its release. The number steadily increased to 1083 by May 2013. By then, our data already recorded a total of 6678 Samsung users, half of whom used the high-end models. Edge numbers generally increased over time, except in February 2013, during which the Lunar New Year was celebrated and people spent more time on family reunions and rest.

A simple empirical data analysis is performed by graphing the data on Samsung Note II adoption with a fitted linear regression curve. This graph facilitates the investigation of the relationship between network structure measures and the diffusion process. Social influence leads to a social multiplier effect within networks, in which a network with social influence has a higher "diffusion speed" than a network with no social influence at a given time period. The difference between the end and

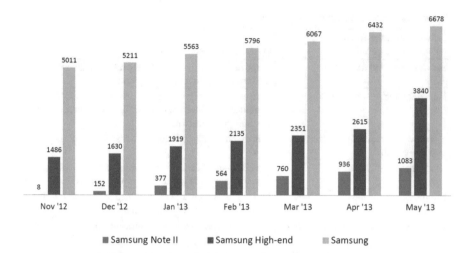

Fig. 12.1 The monthly number of adopters for each level

initial adoption numbers is calculated and divided by the number of time periods to operationalize the concept of diffusion speed. It is used as the y-axis in the graph with respect to different network structure measures in the x-axis; these graphs are shown in Fig. 12.2. Network measures such as network size, density, epidemic threshold, log in/out ratio, time variation of edge numbers, assortativity, diversity, and position of initial adopters, have a clear directional relationship with diffusion speed. The diffusion process is observed to be correlated with the network structure, although these graphs are not obtained from formal statistical tests and the result is biased because other confounding factors are uncontrolled.

12.1 Dynamic Model

Social influence is inferred from correlations of individual behaviors. Social influence is correctly measured by distinguishing it from homophily, which describes the endogenous associations of individuals with similar behaviors or characteristics. The stochastic actor-based model for network dynamics [10–12] is applied to overcome this endogeneity problem. The model estimates the co-evolution of network formation and individual behaviors using longitudinal network information.

12.1.1 Continuous-Time Markov Model Assumption

In our telecommunication data, researchers observe network, g_t, and behavioral outcomes, $y_t = (y_{1t}, \ldots, y_{Ht})$, of n individuals at two or more discrete points in

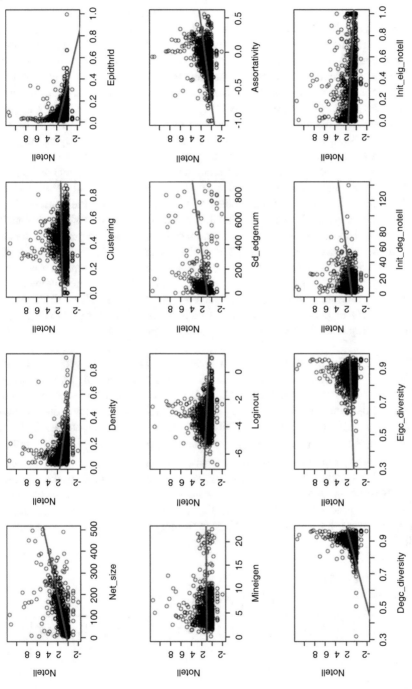

Fig. 12.2 Plots of relationships between variation of Note II adoption and network characteristics

time $t \in \{t_1, \ldots, t_M | t_1 < \ldots < t_M\}$. Network g_t is represented by a $n \times n$ adjacency matrix, with each entry $g_{ij,\,t}$ equal to one if actor i connects to actor j; otherwise, it is equal to zero. Network links in g_t are undirected; thus, g_t is symmetric. Each y_{ht} is a $n \times 1$ vector, and element $y_{i,\,ht}$ is a binary variable that is equal to one if actor i adopts product h; otherwise, it is equal to zero. The stochastic actor-based model views these networks and outcome observations as snapshots from a continuous-time Markov process. The model assumes that there are "micro" steps taking place at stochastically-determined moments between any two discrete time points t_m and t_{m+1}, at which individuals can alter their network ties or behaviors. We let $Z_t = (g_t, y_t)$ denote the state variable; changes in the future state Z_{t+r}, $r > 0$, depend only on the current Z_t because of the Markov chain property.

The model can provide causal interpretations of the homophily effect and social influence by assuming that individual changes in network ties and behaviors are conditionally independent of each other at the given state of the process. The co-evolution of network ties and behaviors are therefore separated into network formation and social influence processes, respectively. While one process is occurring, another process is held constant. The model also assumes that only one network tie or behavioral variable can be changed at a time. This assumption rules out complexity because of the joint action of individuals. Moreover, each network tie or behavior variable could change by only one unit at a time. These assumptions allow changes to occur in small steps, which in turn form a parsimonious and relatively simple Markov process in the study of the co-evaluation of networks and behaviors.

The core of the model is a micro-mechanism in which actors execute myopic changes on network ties and behaviors based on individual utility functions. A Poisson process with the rate function $\lambda_{i,\,m}$ determines when actor i should execute changes in the time period $t_m \leq t \leq t_{m+1}$. For simplicity, we assume that the rate functions of changing either network ties or behaviors are homogeneous for all n members in the same time period (i.e., $\lambda_{i,\,m}^{[g]} = \lambda_m^{[g]}$ and $\lambda_{i,\,m}^{[y_h]} = \lambda_m^{[y_h]}$ for all i). The individual utility function consists of an evaluation function $f_i(\beta, Z)$, which depends only on the current state configuration and on a stochastic error from the extreme type-I distribution. Actor i adjusts one network tie (or one behavior) to optimize his/her utility function. For the network tie, the probability of actor i choosing to change the network tie with actor j, which is from $g_{ij,\,t}$ to $1 - g_{ij,\,t}$, takes the binary logit form,

$$P\left(1 - g_{ij,t}, g_{-ij,t} | Z_t\right)$$

$$= \frac{\exp\left(f_i^{[g]}\left(\beta^{[g]}, 1 - g_{ij,t}, g_{-ij,t}, y_t\right)\right)}{\exp\left(f_i^{[g]}\left(\beta^{[g]}, g_{ij,t} = 1, g_{-ij,t}, y_t\right)\right) + \exp\left(f_i^{[g]}\left(\beta^{[g]}, g_{ij,t} = 0, g_{-ij,t}, y_t\right)\right)}$$

$$(12.1)$$

where $g_{-ij,\,t}$ stands for g_t which excludes $g_{ij,\,t}$. For the behavior, the probability of changing the behavioral variable $y_{i,\,ht}$, which is from $y_{i,\,ht}$ to $1 - y_{i,\,ht}$, is given by

$$
\begin{aligned}
& P\left(1 - y_{i,\,ht}, y_{-i,\,ht} | Z_t\right) \\
& = \frac{\exp\left(f_i^{[y_h]}\left(\beta^{[y_h]}, 1 - y_{i,\,ht}, y_{-i,\,ht}, g_t\right)\right)}{\exp\left(f_i^{[y_h]}\left(\beta^{[y_h]}, y_{i,\,ht} = 1, y_{-i,\,ht}, g_t\right)\right) + \exp\left(\beta^{[y_h]}, y_{i,\,ht} = 0, y_{-i,\,ht}, g_t\right)}
\end{aligned}
\tag{12.2}
$$

where $y_{-i,\,ht}$ stands for y_{ht} which excludes $y_{i,\,ht}$. Here we consider binary behavioral variables, i.e., adopt versus not adopt. In Snijders et al. [12], behavioral variables can be polychotomous. If the stationary transition probability of the state variable Z from this continuous-time Markov chain exists, then it is fully described by the matrix of transition intensities:

$$
q(z, z') = \lim_{d_t \downarrow 0} \frac{P(Z_{t+d_t} = z' | Z_t = z)}{d_t},
\tag{12.3}
$$

where z and z' denote the current and next state, respectively. To be explicit, $q(z, z')$ have the following elements:

$$
q(z, z')
$$

$$
= \begin{cases}
\lambda^{[g]} P\left(1 - g_{ij}, g_{-ij} | Z\right) & \text{if } z' = \left(1 - g_{ij}, g_{-ij}, y\right) \\
\lambda^{[y_h]} P\left(1 - y_{i,\,h}, y_{-i,\,h} | Z\right) & \text{if } z' = \left(1 - y_{i,\,h}, y_{-i,\,h}, g\right) \\
-\sum_i \left\{ \sum_{j \neq i} \lambda^{[g]} P\left(1 - g_{ij}, g_{-ij} | Z\right) + \sum_h \lambda^{[y_h]} P\left(1 - y_{i,\,h}, y_{-i,\,h} | Z\right) \right\} & \text{if } z' = z \\
0 & \text{otherwise}
\end{cases}
\tag{12.4}
$$

In Eqs. (12.1) and (12.2), we specify for $h = 1 \ldots H$:

$$
f_i^{[g]}\left(\beta^{[g]}, Z_t\right) = \sum_l \beta_l^{[g]} s_{il}^{[g]}(Z_t),
\tag{12.5}
$$

$$
f_i^{[y_h]}\left(\beta^{[y_h]}, Z_t\right) = \sum_l \beta_l^{[y_h]} s_{il}^{[y_h]}(Z_t),
\tag{12.6}
$$

to describe the evaluation functions for network ties and behaviors separately. By properly selecting $s_{il}^{[g]}(Z)$ and $s_{il}^{[y_h]}(Z_t)$ in the evaluation functions, the stochastic actor-based model can capture the homophily effect, social influence, and other effects from the network structure and exogenous variables. We specify the "average similarity effect" in the evaluation function of behaviors to capture social influence, which expresses the conformity motive of an individual in behaving as his/her peers do, particularly when enough peers behave in the same manner [4, 16, 17]. For actor i, the average similarity effect on behavior y_h is defined as

$$s_{il}^{[y_h]}(Z) = g_{i+}^{-1} \sum_{j \neq i} g_{ij} \left(sim_{ij}^{[y_h]} - \widehat{sim}^{[y_h]} \right) \qquad (12.7)$$

where $g_{i+} = \sum_{j \neq i} g_{ij}$, $sim_{ij}^{[y_h]} - \widehat{sim}^{[y_h]}$ is the centralized similarity score with $sim_{ij}^{[y_h]}$ $= 1 - | y_{hi} - y_{hj} |$ and $\widehat{sim}^{[y_h]}$ is the mean of similarity scores across all pairs. The parameter $\beta_1^{[y_h]}$ for the average similarity effect reflects the size of the social influence effect. The homophily effect of behavior y_h on the evaluation function of network ties is also captured by the "similarity effect":

$$s_{il}^{[g]}(Z) = \sum_{j \neq i} g_{ij} \left(sim_{ij}^{[y_h]} - \widehat{sim}^{[y_h]} \right) \qquad (12.8)$$

12.1.2   Model Estimation and Identification

The likelihood function of the stochastic actor-based model can be used with the probability structure governed by the continuous-time Markov process to estimate the unknown parameters $\theta = (\lambda, \beta)$ in the rate and evaluation functions. However, the resulting likelihood function will not have a closed form, which renders implementing the maximum likelihood (ML) or Bayesian method difficult. We therefore follow Snijders [11] and Snijders et al. [12] in estimating the model through the method of moments (MoM). We let $\mu(Z)$ denote a vector of statistics based on the state variable Z. MoM estimators are determined by solving the moment equations in which the expected and observed sample statistics resemble each other (i.e., $E_{\hat{\theta}}(\mu(Z)) = \mu(Z)$, where $\mu(Z)$ denotes the observed ones). Given that the expected sample statistics cannot be calculated explicitly, they are replaced with the averages of the statistics from the simulated samples. According to the micro-steps of the network and behavior evolutions outlined in Sect. 5.1 the changes in network ties and behaviors between any two network observations are simulated by conditioning the initial configuration of Z.

Vector $\mu(Z)$ has the same dimension as parameter vector θ, which should be chosen so that a corresponding element in $\mu(Z)$ sensitively responds to changes in each separated parameter in θ. Substantial variations in network ties and behaviors over time, which are reflected in $\mu(Z)$, are necessary for parameter identification. The rate parameters (i.e., $\lambda_m^{[g]}$ and $\lambda_m^{[y_h]}$) influence only the Poisson process between t_m tm and t_{m+1}. Thus, the moment equation that we employ is as follows:

$$E_{\theta}\{\mu_m(Z_{t_m}, Z_{t_m+1}) | Z_{t_m} = z_{t_m}\} = \mu_m(z_{t_m}, z_{t_m+1}) \qquad (12.9)$$

and the choice of $\mu_m(Z_{t_m}, Z_{t_m+1})$ in Eq. (12.9) are $\sum_{ij} |g_{ij,t_{m+1}} - g_{ij,t_m}|$ and $\sum_{ij}$ $|y_{hi,t_{m+1}} - y_{hi,t_m}|$ for estimating $\lambda_m^{[g]}$ and $\lambda_m^{[y_h]}$, respectively. The parameters β in the evaluation functions are constant and appear in all of the statistics $\mu_m\left(Z_{t_m}, Z_{t_{m+1}}\right)$, $m = 1, \ldots, M - 1$. Thus, the moment equation used is:

$$\sum_{m=1}^{M-1} E_\theta\left\{\mu_m\left(Z_{t_m}, Z_{t_{m+1}} | Z_{t_m} = z_{t_m}\right)\right\} = \sum_{m=1}^{M-1} \mu_m\left(z_{t_m}, z_{t_{m+1}}\right), \tag{12.10}$$

and the sample statistics used to estimate $\beta^{[g]}$ and $\beta^{[y_h]}$ are based on $\sum_{i=1}^{n} s_{il}(Z)$ with $s_{il}(Z)$ from Eqs. (12.8) and (12.7). Increasing $\beta_l^{[g]}$ (or $\beta_l^{[y_h]}$) gives a larger influence of similarity effect on individuals evaluation function, which causes a higher chance of forming link (or adopting product) and leads to a higher similarity effect on all the actors in the subsequent time moments. Although these two sample statistics respond well to the change in $\beta^{[g]}$ and $\beta^{[y_h]}$, they are perfectly multicollinear and produce two identical moment equations. To prevent underidentification, we follow Snijders et al. [12] to exploit the time order of variables following the concept of causality. Homophily is reflected as a "later" change in the network tie, following the "earlier" configuration of behaviors, while social influence is reflected as a "later" change in behaviors following the "earlier" configuration of network ties. The homophily effect on network formation is estimated using the following sample statistic:

$$\mu_m(Z_m, Z_{m+1}) = \sum_i s_{il}^{[g]}\left(g_{t_{m+1}}, y_{t_m}\right). \tag{12.11}$$

The social influence effect on behavior y_h is estimated using

$$\mu_m(Z_m, Z_{m+1}) = \sum_i s_{il}^{[y_h]}\left(y_{h, t_{m+1}}, y_{-h, t_m}, g_{t_m}\right), \tag{12.12}$$

where y_{-h, t_m}, denotes y_{t_m} which excludes y_{h, t_m}. When the MoM estimator $\hat{\theta}$ is obtained, the delta method is used to calculate the approximate covariance matrix of $\hat{\theta}$.

12.1.3 Meta-Analysis of Network Structure Effects on Social Influence

For a systematic overview of the results from our large number of network samples, metaregression analysis is applied to identify the association between the effects of network structure and social influence. Dependent variables include both the binary indicator of significance and the magnitude of the social influence effect on cellphone adoptions in three levels of choice hierarchy. The indicator of significance, which represents the existence of social influence, is studied using the standard binary probit model. The magnitude of social influence, which represents the size of the social influence effect, is studied using the random effect meta-regression model to incorporate both within- and between-network heterogeneities. We let S_i and σ_i^2 denote the estimated social influence effect and the corresponding variance from network sample i. Based on the random effect assumption, the true social influence effect θ_i from sample i follows a normal distribution centered at linear predictor $x_i\beta$:

$$S_i \mid \theta_i \sim N\left(\theta_i, \sigma_i^2\right), \text{where} \ \ \theta_i \tilde{} N\left(x_i\beta, \tau_i^2\right), \tag{12.13}$$

where x_i denotes a vector of network characteristics. The model can be rewritten as

$$S_i = x_i\beta + u_i + \varepsilon_i, \text{where} \ \ u_i \sim N\left(0, \tau^2\right) \ \text{and} \ \ \varepsilon_i \tilde{} N\left(0, \sigma_i^2\right), \tag{12.14}$$

The weighted least square method is used to estimate the unknown slope coefficients β and the between-network variance τ^2 in Eq. (12.14); τ^2 is estimated using the restricted maximum likelihood (REML) method [18].

12.2   Summary of Findings

We had three main findings, which can be summarized as follows. First, we estimated the effect of social influence and homophily from cellphone adoption sample and studied the effect of demographic characteristics on the existence and magnitude of social influence. Regarding the adoption behaviour, 6.0% of approximately 1000 individual social networks exhibit social influence for Samsung Note II adoption, 12.3% for Samsung high-end phone adoption, and 10.2% for Samsung brand adoption. Despite the fact that adoption rate for Samsung branded phones is higher than that for the Samsung high-end phones, we see higher social influence in high-end phone adoption. A reason for this is that social influence is determined through the variation in the adoption numbers, that is, through the concept of flow instead of stock. No monotone relationship therefore holds between absolute adoption rate and social influence.

These counterintuitive results demonstrate the value of the current analysis. We find significant and positive social influence effects in all three behavioral cases. The effect implies that if all the friends of an individual who has not yet adopted the cellphone product have already adopted the said product, then social influence increases the chances of that individual adopting the product by 7.38 times. This highlights the importance of network information. Our second finding focuses on examining the relationship between various network characteristics and social influence. We have found that, all measures, excluding minimum eigenvalue and log in-out edge ratio, are significant predictors of social influence. Diversity of connection (network structural entropy) and time variation of edge numbers are the two most important network measures related to the social influence effect. This indicates that evenly distributed or expanding networks have high social influence effects. Also, social influence mostly occurs in networks with large-sized nodes and high-density links, regardless of choice hierarchy. The number of first- and second-degree friends drives diffusion [6], and our results confirm that it also influences the process. Lastly, a high average number of links for each individual (low threshold) causes the network to exhibit a large social influence effect. This is especially apparent when a well-connected individual interacts with another well-connected individual. The initial adopter status also significantly affects the diffusion process.

Our third finding from the policy simulation had provided insight on seeding strategy. It appears that the social influence effect is a double-edged sword. More specifically, when the individual has contact with many adopters, the social influence effect increases the probability of that individual becoming an adopter. However, when the individual meets only a few adopters in the network, the social influence effect decreases the probability of that individual becoming an adopter. Members in networks with social influence may become more reluctant to adopt a new model other than the "mainstream model" that is already adopted in their network. A certain number of adopters therefore are needed to enable the diffusion process. Therefore, promotion of the Samsung Note II to increase its adoption will likely be difficult with the existence of social influence; individuals are more likely to adopt Samsung branded phones, particularly high-tier ones due to the promotion. When evaluating the effectiveness of a promotion seeding strategy, relying solely on the adoption rate of the new product is not sufficient. Its contribution to the related brand and category should also be considered.

12.2.1   Estimation Results from SIENA

We rely on the variations in network links and behaviors across time to identify the parameters of the rate and evaluation functions in the stochastic actor-based model. Identification in certain networks and behaviors with infrequent changes is weak, and the estimation algorithm may not converge. This issue threatens the behavioral variable of Samsung Note II adoption because it only accounts for 7% within our sample period. The three adoption variables are separated into three models to minimize the chain effect of non-convergence caused by the weak identification of the social influence effect from one behavioral variable to other estimates; we independently estimate these three models.

Figure 12.3 provides the scatter plots of the estimated social influence and homophily effects obtained from the stochastic actor-based model. Each dot in the scatter plot represents an estimate of the effect (in the horizontal axis) and the corresponding standard error (in the vertical axis) from one network sample. We include only networks that exhibit converged estimates of the social influence effect, which left us with 410 networks for Samsung Note II, 715 for Samsung high-end models, and 791 for Samsung brand adoptions. These networks are separated into significant and insignificant points based on the t-ratio of significance at 10% level. Figure 12.3a shows that the converged estimates of the social influence effect range from 2.5 to 5. The significant social influence effects are all positive in the three

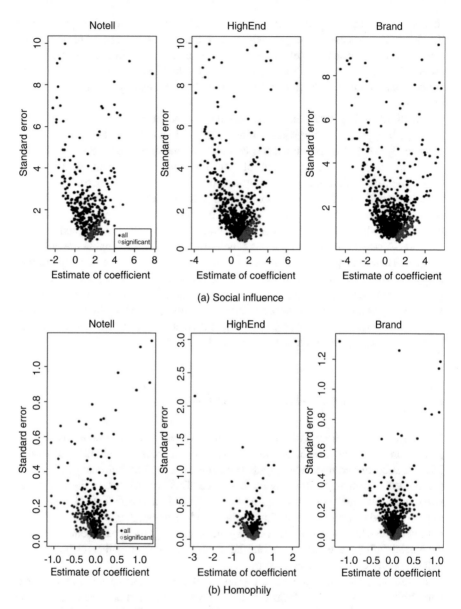

Fig. 12.3 Scatter plot of estimated social influence

behavioral cases with a mean of approximately 2. If an individual has not yet adopted the cellphone product when all of his/her friends have already adopted it, then social influence increases the likelihood of the individual adopting the product by exp (2) = 7.38 times higher than the case that the individual makes no change. We take the same groups of networks used in Fig. 12.3a and present the estimate and standard error of the homophily effect in Fig. 12.3b. The scatter plots show that the converged estimates of the homophily effect can either be positive or negative. Significant estimates for the homophily effect are not particularly different and also range from negative to positive. Therefore, we cannot claim the general conclusion that the homophily effect of cellphone adoption exists for network formation; no pervasive pattern shows that individuals become friends because of adopting Samsung Note II.

12.2.2 Results of Meta-Regression Analysis

We conduct meta-regression analysis to identify the types of network characteristics that serve as predictors of social influence effects. The network samples used in this analysis are those with converged estimates of social influence effects, as shown in Fig. 12.3. Table 12.2 provides the summary statistics of variables used in the regressions. The accumulated network is employed across sampling periods to construct the network statistics. The sex ratio (percentage of males) is approximately 60% in our network samples. The average age of the customers is 37, and their average tenure (customership) is 33 months. For social influence effects, 13.6% of the networks with converged results have significant estimates and an average magnitude of the estimates for Samsung Note II adoption is 0.964; 16.2% of the networks have significant estimates and an average magnitude for Samsung high-end adoption is 0.798; and 12.1% of the networks have significant estimates with an average magnitude of 0.820 for Samsung brand adoption. Although the number of adoptions at the low levels of the choice hierarchy is always higher than that at the high levels, the significance and magnitude of the estimated social influence effect do not follow the same order across hierarchies. According to our identification strategy, social influence effects are identified from variations in adoption behaviors over time (i.e., the concept of "flow" rather than the absolute adoption number or the "stock"). Figure 12.1 shows that high-end adoption significantly increased relative to brand adoption during the sample period. Significant social influence effects are therefore exhibited more in high-end adoptions than in brand adoptions.

In this study, we have encountered the multicollinearity problem due to the high correlations among network statistics that were incorporated into the same regression model. This is solved by separating the network statistics, and only one network statistic is added into the benchmark model (where network size and density are controlled) each time to study its partial effect. Separating network characteristics in different regressions facilitates the interpretation of the influence by preventing the influences of network statistics from confounding each other. The benchmark model also controls for customer age, sex, and tenure to examine how demographic variables affect social influence.

Table 12.3 presents the estimation results of the benchmark model. The average age of the customers negatively affects on social influence, particularly for Samsung high-end and brand adoptions. In other words, the youth are likely to succumb to

Table 12.2 Summary statistics of variables used in meta regression

	Note II		High End		Brand		Pool	
	Mean	Std. Dev.	Mean	Std. Dev.	Mean	Std. Dev.	Min	Max
Social influence								
Significance	0.136	0.343	0.162	0.368	0.121	0.326	0.000	1.000
Magnitude	0.964	3.461	0.798	2.825	0.820	2.953	−46.973	50.211
Demography								
Sex ratio	0.606	0.149	0.621	0.136	0.619	0.141	0.000	1.000
Age	37.728	3.302	37.740	3.326	37.705	3.205	21.373	59.933
Tenure	32.938	12.794	33.733	10.943	33.849	10.627	0.000	63.000
Network measures								
Net size	90.917	86.827	100.189	88.197	95.220	85.187	4.000	495.000
Density	0.154	0.101	0.143	0.108	0.141	0.103	0.017	0.900
Clustering coefficient	0.445	0.157	0.433	0.151	0.427	0.154	0.000	.875
Minimum eigenvalue	5.579	3.893	5.548	3.658	5.508	3.618	0.000	22.437
Log in-out edge ratio	−3.334	1.092	−3.331	1.078	−3.287	1.110	−5.979	0.587
Epidemic threshold	0.105	0.077	0.104	0.079	0.106	0.077	0.009	0.500
Assortativity	−0.086	0.233	−0.098	0.227	−0.112	0.232	−1.000	0.551
Degc. diverstiy	0.899	0.038	0.897	0.039	0.895	0.040	0.700	0.969
Eigc. diversity	0.863	0.058	0.859	0.060	0.857	0.060	0.627	0.969
Init adopter degc.	10.591	13.072	10.341	9.515	8.762	8.590	0.000	139.000
Init adopter eigc.	0.321	0.282	0.321	0.212	0.276	0.163	0.000	1.000
Sd of edge number	66.821	139.354	62.240	135.585	52.783	118.475	0.756	882.283
Sample size	410		715		791		856	

Note: The samples in the last column "pool" are the union of samples from the first three columns. Degc. stands for degree centrality. Eigc. stands for eigenvector centrality

Table 12.3 Meta regression of social influence – benchmark case based on network size and density

	Samsung Note II		Samsung HighEnd		Samsung brand	
	Significance	Magnitude	Significance	Magnitude	Significance	Magnitude
Net size	0.00827*** (8.17)	0.00154* (2.17)	0.00905*** (11.21)	0.00256*** (6.19)	0.00924*** (11.06)	0.00269*** (6.40)
Density	3.057*** (3.75)	3.866*** (3.42)	2.654*** (3.97)	5.118*** (7.39)	3.383*** (5.27)	4.785*** (7.46)
Sex ratio	−0.430 (−0.65)	1.029 (1.27)	−0.530 (−1.01)	−0.210 (−0.45)	−0.530 (−1.04)	0.166 (0.39)
Age	0.000328 (0.01)	−0.0365 (−1.10)	−0.0606* (−2.58)	−0.0404* (−2.02)	−0.0755** (−3.01)	−0.0507** (−2.74)
Tenure	0.0168 (1.35)	−0.00151 (−0.12)	0.0113 (1.21)	−0.00869 (−1.24)	0.0238* (2.43)	0.00916 (1.27)
Constant	−2.876* (−2.39)	1.487 (1.20)	−0.306 (−0.35)	2.008** (2.61)	−0.500 (−0.53)	1.402+ (1.94)
R^2	0.247	0.051	0.259	0.187	0.283	0.184
τ^2	–	0.00	–	0.00	–	0.00
Observations	410		715		791	

Note: t statistics in parentheses. The variable significance is studied by a binary Probit model. The variable magnitude is studied by the (random) mixed effect meta regression model. The estimates of between-network variance, τ^2, are zero for all three cases. Pseudo (adjusted) R^2 are reported for the dependent variable of significance (magnitude). $^{+}p < 0.10$, $^{*}p < 0.05$, $^{**}p < 0.01$, $^{***}p < 0.001$

social influence. The effect of customer tenure on social influence is limited to the case of adopting at the Samsung brand level. The sex ratio generally has no significant effect on social influence. Both large size and high density significantly contribute to the existence and high magnitude of social influence within networks, which applies to adoption behaviors in all three levels. Our network size is endogenously determined by two layers of snowball sampling, and our results echo the findings of Yoganarasimhan [6] on the first- and second-degree connectivity of seeds in video dissemination. The pervasive homophily effect driven by product adoption has not been observed in this study, meaning individuals do not connect with one another just because they have the same phone.

Table 12.4 presents the results of the separate regressions. The REML estimates for $\tau 2$ are zero (i.e., no between-network heterogeneity) in all meta-regressions for social influence magnitudes. The results can be therefore interpreted as the fixed effect meta-regression estimates obtained from the weighted least square method, with a weight of $1/\sigma i 2$ for each network i. We first observe the effect of the clustering coefficient on social influence. The clustering coefficient has an insignificant effect on the social influence for Samsung Note II adoption but a positive and significant effect on the social influence for Samsung high-end and brand adoption cases. The closure of network ties may be unnecessary for stimulating the social influence for Samsung Note II adoption. However, if the friends of an individual are friends with one another, then their adoption of any Samsung or Samsung high-end smartphones will also influence the individual adopting a Samsung or Samsung high-end smart phones.

The minimum eigenvalue of a network matrix captures the extent to which the network amplifies the substitutabilities of agent actions [14]. The estimation result shows that the absolute minimum eigenvalue cannot predict the social influences for cellphone adoptions at any level. One explanation could be because our network samples are far from the "two-side," which are needed to have a great rebound of actions. The estimates for epidemic threshold are negative and significant at 5% level for all cases, which implies that a high (low) epidemic threshold results in a weak (strong) social influence. We also examine the assortativity effect (i.e., correlation patterns of degrees of connected nodes) on social influence. The estimated coefficients for assortativity are positive and significant for high-end and brand adoptions but insignificant for Samsung Note II adoption. We then examine the effect of log inward-outward edge ratio on social influence. The estimated effects of log in/out ratio on social influence are positive but insignificant in all cases; this finding is in agreement with the theoretical prediction of Young [13], which states that strong inward connections in local clusters relative to outward connections facilitate social influence. Previous empirical studies have shown the contradicting results of this effect on adoption behavior. For example, Yoganarasimhan [6] suggested a negative global effect, whereas Katona et al. [5] argued a positive local effect. Therefore, the effects are reconciled and become insignificant when looking at process. The discrepancies highlight the uniqueness of the current analysis to complement our understanding of diffusion and network structure.

Table 12.4 Meta regression of social influence – collection of results from separated regressions

	Samsung Note II		Samsung HighEnd		Samsung brand	
	Significance	Magnitude	Significance	Magnitude	Significance	Magnitude
Clustering coefficient	−0.175 (−0.23, 0.247)	0.222 (0.28, 0.049)	2.171** (3.25, 0.276)	1.258*** (3.79, 0.202)	3.994*** (5.29, 0.337)	2.168*** (7.37, 0.236)
Minimum eigenvalue	0.0212 (1.01, 0.249)	−0.0074 (−0.50, 0.049)	0.0212 (1.28, 0.261)	0.0054 (0.65, 0.187)	−0.0132 (−0.66, 0.284)	−0.0031 (−0.35, 0.184)
Epidemic threshold	−7.437** (−3.22, 0.263)	−5.872*** (−3.42, 0.094)	−18.74*** (−5.08, 0.286)	−6.488*** (−5.40, 0.254)	−8.581*** (−4.57, 0.312)	−5.546*** (−5.26, 0.239)
Assortativity	0.542 (1.12, 0.250)	0.646 (1.24, 0.055)	1.457*** (3.81, 0.282)	0.860** (2.92, 0.206)	1.579*** (4.12, 0.312)	1.258*** (4.63, 0.227)
Log in-out edge ratio	0.0130 (0.12, 0.243)	0.0453 (0.39, 0.049)	0.0319 (0.39, 0.258)	0.0772 (1.16, 0.190)	0.0481 (0.60, 0.285)	0.0299 (0.51, 0.189)
Degc. diversity	13.04*** (3.78, 0.294)	8.326* (2.50, 0.073)	13.24*** (4.66, 0.300)	7.230*** (3.70, 0.218)	10.29*** (3.95, 0.313)	7.269*** (4.10, 0.217)
Eigc. diversity	5.117** (3.05, 0.275)	2.683* (2.07, 0.059)	3.149** (2.65, 0.270)	1.816** (2.94, 0.196)	2.055 (1.64, 0.288)	0.562 (1.00, 0.184)
Initial adopters' degc.	0.0113+ (1.89, 0.257)	0.00478 (1.04, 0.053)	0.0302*** (3.82, 0.282)	0.00593 (1.24, 0.190)	0.0251 (2.86, 0.297)	0.00941 (1.60, 0.189)
Initial adopters' eigc.	−0.417 (−1.07, 0.250)	−0.349 (−0.96, 0.052)	−0.727 (−1.60, 0.263)	−0.681+ (−1.92, 0.195)	−2.080** (−3.07, 0.299)	−1.556** (−3.17, 0.204)
Sd of edge number	0.00213*** (3.61, 0.286)	0.0023*** (5.03, 0.105)	0.000475 (1.10, 0.260)	0.0012*** (4.93, 0.213)	0.000509 (1.03, 0.285)	0.0012*** (3.67, 0.197)
Observations	410		715		791	

Note: The results reported in this table are collected from several separated regressions with each one focusing on one specific network statistic. In all of the separated regressions, we control network size, network density, and customer demographics. Values reported in parentheses are t statistics of the coefficients (left) and R^2's of the regressions (right). + $p < 0.10$, * $p < 0.05$, ** $p < 0.01$, *** $p < 0.001$.

The estimated coefficients for the structural entropy based on degree centrality are positive and significant at 5% level for all cases. The estimated coefficients for the structural entropy based on eigenvector centrality are also positive but only significant for Samsung Note II and high-end adoptions. A comparison of the diversities of the degree and eigenvector centrality shows that the diversity of local connections is more important than the diversity of global connections in facilitating social influence. The average degree centrality of initial adopters shows a positive and significant effect on social influence at 5% level, which supports our hypothesis. However, the effect of the average eigenvector centrality of initial adopters is negative and significant at 5% level. The eigenvector centrality shows the importance of nodes connectors. If an initial adopter has high eigenvector centrality, this means that he/she is surrounded by other active individuals in a network, which in turn marginalizes his/her influence. Finally, edge number variation has a positive effect on social influence in all cases and is particularly significant for Samsung Note II adoption but less significant for high-end and brand adoptions. The variation in edge numbers is generally positive; thus, expanding networks is more likely to exhibit social influence.

A comparison of all these network statistics demonstrated that the diversity of local connections and variation of edge numbers are the two most important effects that facilitate social influence, based on the significance of the effects and corresponding $R2$ in each separate regression. Thus, we pool these two effects along with network size and density in one additional meta-regression. These four network effects remain positive and significant for most cases. $R2$ is significantly improved relative to the benchmark model in Table 12.3, particularly for the social influence magnitude on Samsung Note II adoption.

12.2.3   Policy Simulation

In a policy simulation based on our model, we consider a scenario in which the wireless carrier relies on the social influence of telecommunication networks to promote the adoption of Samsung cellphone products. In the first step, the company provided free products or price discounts to a number of customers selected from the networks. These initial users stimulate and propagate the social influence on product adoption in their own social networks. We selected two real networks from our sample for the simulation. These two networks (No. 838 and No. 301) have similar sizes but different characteristics. Table 12.5 and Figs. 12.4 and 12.5 show that No. 838 is denser and has a larger time variation in edge numbers than No. 301. A comparison of the simulation results from these two networks provides useful information for selecting networks for product promotion.

The simulation starts with a seeding strategy in selecting initial adopters in the network. Three seeding strategies are considered: purely random, targeting individuals with high degree centralities, and targeting individuals with high eigenvector centralities. The simulation is iterated for 500 rounds. At each round r, r = 1...500,

Table 12.5 Network characteristics of two selected networks in simulation

Network	301	838
Social influence		
NoteII	−10.6178 (1018.51)	1.7964 (0.7235)
HighEnd	0.9671 (0.6251)	1.8555 (0.5839)
Brand	0.5409 (0.6158)	1.2688 (0.6621)
Size	329	330
Density	0.0354	0.1644
Clustering coefficient	0.3140	0.4560
Minimum eigenvalue	−4.0110	−3.5233
Sd of edge number	60.3896	318.9021
Epidemic threshold	0.0380	0.0111
Assortativity	0.0747	0.0287
Degc. diversity	0.9136	0.9436
Eigc. diversity	0.8422	0.9390
Initial Note II adopters' degc.	21.000	31.400
Initial high-end adopters' degc.	15.232	62.166
Initial brand adopters' degc.	13.235	60.813
Initial Note II adopters' Eigc.	0.3533	0.2094
Initial high-end adopters' Eigc.	0.2099	0.3970
Initial brand adopters' Eigc.	0.1809	0.3894

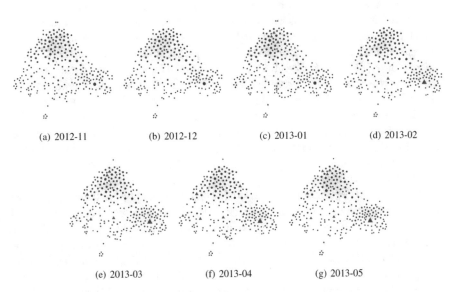

(a) 2012-11   (b) 2012-12   (c) 2013-01   (d) 2013-02

(e) 2013-03   (f) 2013-04   (g) 2013-05

Fig. 12.4 Evolution of network ties and adoption in network No. 301

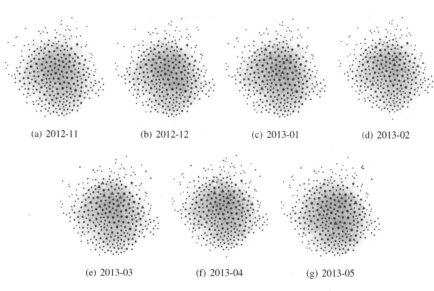

(a) 2012-11   (b) 2012-12   (c) 2013-01   (d) 2013-02

(e) 2013-03   (f) 2013-04   (g) 2013-05

Fig. 12.5 Evolution of network ties and adoption in network No. 838

where all the individuals in a network have one chance to change their adoption status following a random order. The logit probability function of Eq. (12.2) determines if an individual changes status from 0 (non-adopter) to 1 (adopter) with a designed constant to reflect a baseline adoption rate. The social influence effect is represented in this logit probability function by the average similarity effect estimated from the stochastic actor-based model. Network No. 838 has significant average similarity effects for all three Samsung cellphone choice levels with coefficients of 1.796, 1.855, and 1.228, respectively, whereas none of the average similarity effects in No. 301 are significant (Table 12.5). Thus, social influences in No. 301 do not exist. Once an individual becomes an adopter, their status will never revert. The simulation is repeated for 100 times, and the average number (with the 95% confidence interval) of additional adopters for three adoption choices are plotted in Fig. 12.6, in which we report the results for the random seeding and the high degree centralities seeding. The results for the high eigenvector centralities seeding are similar to that of the high degree centralities seeding,

Based on the simulations, we find that social influence in the form of the similarity effect is a double-edged sword. When an individual has contact with several adopters, the similarity effect increases his/her probability of becoming an adopter. However, when an individual has contact with only a few adopters in a network, the similarity effect discourages this individual from adoption. The top row in Fig. 12.6 shows that when there are only 50 initial Samsung Note II adopters in the network, the social influence effect discourages adoption in No. 838, which leads to a lower adoption number than that in No. 301 (with no social influence effect) When Samsung Note II adoption has only 10 initial adopters in the network, the difference

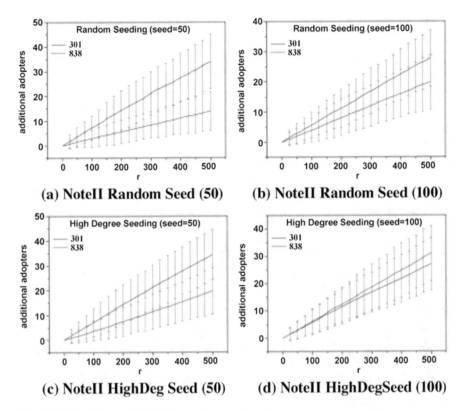

Fig. 12.6 Simulation results for Samsung Note II adoption

is even more pronounced. When the number of initial adopters is further increased to 100, No. 838 eventually gains more Samsung Note II adopters than No. 301.

A comparison of the different seeding strategies shows that choosing individuals with high degree centralities as initial adopters effectively promotes Samsung Note II adoption, whereas randomly choosing initial adopters does the worst among the three. The second and third rows in Fig. 12.6 show the simulation results for Samsung high-end and brand adoptions. The scenarios of either 50 or 100 initial high-end and brand adopters are reasonable targets because they are close to the true market share. In our sample, the market share of Samsung cellphones is 32.3%. We observe similar adoption patterns to Samsung Note II adoption; No. 838 has more additional adopters than No. 301 in the group with 100 initial adopters. In summary, carriers will have a difficult time increasing adoption for Samsung Note II in networks with social influence. However, this promotion will make people more likely to adopt some model of Samsung phones, particularly high-tier ones. This crossover effect on adoptions implies that when evaluating the effectiveness of a promotion seeding strategy, using only the adoption rate of the new product is not sufficient. Its contribution to the related brand and category should also be considered.

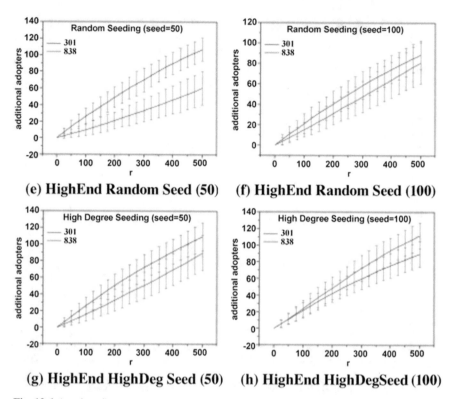

(e) HighEnd Random Seed (50)   (f) HighEnd Random Seed (100)

(g) HighEnd HighDeg Seed (50)   (h) HighEnd HighDegSeed (100)

Fig. 12.6 (continued)

12.3 Conclusion

In this chapter, we have studied the effect of dynamic network structure on social influence, using multiple network measures at both micro and macro levels. This study contributes to existing research in a few ways. First, we use various network measures to understand the diffusion process, particularly social influence, and at different levels of choice hierarchy. Second, we propose the application of a new model in analyzing co-evolution of network formation and behavior, and simultaneously observe the social influence effect. Lastly, our findings provide companies with new means of conducting buzz marketing campaigns through the use of simple network structure measures to select the right local network as the starting point of the seeding strategy.

Diffusion of a Samsung smartphone was examined within a major wireless carrier across two medium-sized cities in China. The social influence in each social network is identified and measured using the stochastic actor-based model. Meta-analysis comparing networks had revealed two important network measures as the major drivers of social influence, which are diversity of network connections and the standard deviation of edge numbers over time. Thus, the social influence effect is

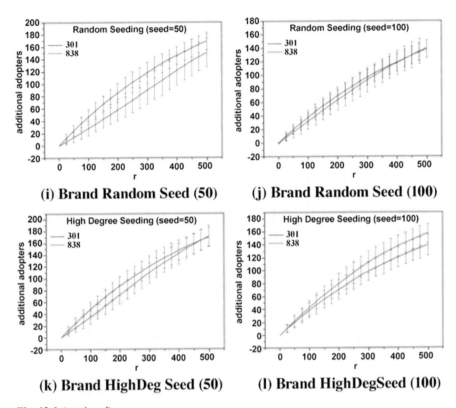

(i) Brand Random Seed (50)   **(j) Brand Random Seed (100)**

(k) Brand HighDeg Seed (50)   **(l) Brand HighDegSeed (100)**

Fig. 12.6 (continued)

likely to occur in evenly distributed or expanding networks. Other factors which contribute to the occurrence of social influence in networks include large network size and high density, abundant closure, low epidemic threshold, high assortativity, and high centrality of initial adopters when a new product is introduced.

Policy simulation had demonstrated that the social influence effect is a double-edged sword, where a certain amount of initial or existing adopters is necessary to trigger positive social influence during the diffusion process. Promoting Samsung Note II adoption through social influence will be surely difficult in a network with a majority of individuals using other brands. We therefore highlight the importance of choosing the right networks when introducing new products. When evaluating the effectiveness of a promotion seeding strategy, its contribution to the related brand and category should also be considered in addition to the adoption rate of the new product.

The findings of this study provide a new perspective on company seeding strategies. Network structure information significantly complements individual demographic and local position information. The right networks must be selected based on a general network structure prior to selecting the right seeds based on personal characteristics and network positions. The limitation of this study lies in the lack of data. We do not have data on individual mobile networks beyond this carrier, which hinders our ability to capture the social influence at work in the entire social network. With more detailed contract records, there exists the probability of expanding the current analysis by considering different influences from various individuals.

References

1. Hartmann, W. R., Manchanda, P., Nair, H., Bothner, M., Dodds, P., Godes, D., Hosanagar, K., and Tucker, C. (2008). Modeling social interactions: Identification, empirical methods and policy implications. Marketing letters, 19(3–4), 287–304.
2. Watts, D. J. and Dodds, P. S. (2007). Influentials, networks, and public opinion formation. Journal of consumer research, 34(4), 441–458.
3. Eagle, N., Macy, M., and Claxton, R. (2010). Network diversity and economic development. Science, 328(5981), 1029–1031.
4. Young, H. P. (2009a). Innovation diffusion in heterogeneous populations: Contagion, social influence, and social learning. The American economic review, 99(5), 1899–1924.
5. Katona, Z., Zubcsek, P. P., and Sarvary, M. (2011). Network effects and personal influences: The diffusion of an online social network. Journal of Marketing Research, 48(3), 425–443.
6. Yoganarasimhan, H. (2012). Impact of social network structure on content propagation: A study using youtube data. Quantitative Marketing and Economics, 10(1), 111–150.
7. Peres, R. (2014). The impact of network characteristics on the diffusion of innovations. Physica A: Statistical Mechanics and its Applications, 402, 330–343.
8. Chen, X., Chen, Y., and Xiao, P. (2013). The impact of sampling and network topology on the estimation of social intercorrelations. Journal of Marketing Research, 50(1), 95–110.
9. Aral, S., Muchnik, L., and Sundararajan, A. (2009). Distinguishing influence-based contagion from homophily-driven diffusion in dynamic networks. Proceedings of the National Academy of Sciences, 106(51), 21544–21549.
10. Snijders, T. A. (1996). Stochastic actor-oriented models for network change. Journal of mathematical sociology, 21(1–2), 149–172.
11. Snijders, T. A. (2001). The statistical evaluation of social network dynamics. Sociological methodology, 31(1), 361–395.
12. Snijders, T., Steglich, C., and Schweinberger, M. (2007). Modeling the coevolution of networks and behavior. In H. O. Kees van Montfort and A. Satorra, editors, Longitudinal models in the behavioral and related sciences, pages 41–71. Lawrence Erlbaum.
13. Young, H. P. (2011). The dynamics of social innovation. Proceedings of the National Academy of Sciences, 108(Supplement 4), 21285–21291.
14. Bramoull'e, Y., Kranton, R., and D'Amours, M. (2014). Strategic interaction and networks. The American Economic Review, 104(3), 898–930.
15. Eagle, N., Pentland, A. S., and Lazer, D. (2009). Inferring friendship network structure by using mobile phone data. Proceedings of the National Academy of Sciences, 106(36), 15274–15278.
16. L'opez-Pintado, D. (2008). Diffusion in complex social networks. Games and Economic Behavior, 62(2), 573–590.
17. Young, H. P. (2009b). Innovation diffusion in heterogeneous populations: Contagion, social influence, and social learning. The American economic review, 99(5), 1899–1924.
18. Morris, C. N. (1983). Parametric empirical bayes inference: theory and applications. Journal of the American Statistical Association, 78(381), 47–55.

Printed in the United States
By Bookmasters